WEYERHAEUSER ENVIRONMENTAL CLASSICS

William Cronon, Editor

WEYERHAEUSER ENVIRONMENTAL CLASSICS
are reprinted editions of key works that explore human relationships with natural environments in all their variety and complexity. Drawn from many disciplines, they examine how natural systems affect human communities, how people affect the environments of which they are a part, and how different cultural conceptions of nature powerfully shape our sense of the world around us. These are books about the environment that continue to offer profound insights about the human place in nature.

The Great Columbia Plain:
A Historical Geography, 1805–1910
by D. W. Meinig

Mountain Gloom and Mountain Glory:
The Development of the Aesthetics of the Infinite
by Marjorie Hope Nicolson

Tutira: The Story of a New Zealand Sheep Station
by H. Guthrie-Smith

A Symbol of Wilderness:
Echo Park and the American Conservation Movement
by Mark W. T. Harvey

Man and Nature: Or, Physical Geography
as Modified by Human Action
by George Perkins Marsh; edited by David Lowenthal

Conservation in the Progressive Era: Classic Texts
edited by David Stradling

Weyerhaeuser Environmental Classics is a subseries within Weyerhaeuser Environmental Books, under the general editorship of William Cronon. A complete listing of the series appears at the end of this book.

Conservation

in the Progressive Era

CLASSIC TEXTS

Edited by David Stradling

Foreword by William Cronon

UNIVERSITY OF WASHINGTON PRESS

Seattle and London

Conservation in the Progressive Era: Classic Texts
has been published with the assistance of a grant
from the Weyerhaeuser Environmental Books Endowment,
established by the Weyerhaeuser Company Foundation,
members of the Weyerhaeuser family,
and Janet and Jack Creighton.

University of Washington Press
PO Box 50096, Seattle, WA 98145, U.S.A.
www.washington.edu/uwpress

Cataloging-in-Publication Data available
from the Library of Congress

The paper used in this publication is acid-free and recycled from 10 percent
post-consumer and at least 50 percent pre-consumer waste. It meets the
minimum requirements of American National Standard for Information
Sciences—Permanence of Paper for Printed Library Materials,
ANSI Z39.48-1984. ♾ ♲

CONTENTS

Foreword   vii

Preface   xi

Introduction   3

Part 1

DEFINING AND DEBATING CONSERVATION   17

Gifford Pinchot, "Principles of Conservation"   19

Theodore Roosevelt, "Special Message from the
President of the United States"   23

William E. Smythe, "The Miracle of Irrigation"   28

Ladies' Home Journal, "What Is Meant by Conservation?"   32

George L. Knapp, "The Other Side of Conservation"   35

H.J.M. Mattes, "Another National Blunder"   39

Part 2

PERSPECTIVES ON WILDLIFE CONSERVATION   43

George Bird Grinnell, "American Game Protection: A Sketch"   45

Mabel Osgood Wright, "Keep on Pedaling!"   49

William T. Hornaday, *Our Vanishing Wild Life:
Its Extermination and Preservation*   51

David Shepard Merrill, "The Education of a Young Pioneer
in the Northern Adirondacks"   53

Part 3

THE UTILITY OF "CONSERVATION" 57

Samuel Gompers, "Conservation of Our Natural Resources" 59

J. Horace McFarland, "Shall We Have Ugly Conservation?" 61

Mary Ritter Beard, "Civic Improvement" 64

Irving Fisher, "National Vitality, Its Wastes and Conservation" 66

Ellen H. Richards, *Conservation by Sanitation: Air and Water Supply, Disposal of Waste* 70

Part 4

SMOKE AND CONSERVATION IN THE CITY 73

Charles A. L. Reed, "An Address on the Smoke Problem" 75

Mrs. Ernest R. Kroeger, "Smoke Abatement in St. Louis" 78

Herbert M. Wilson, "The Cure for the Smoke Evil" 80

Ernest L. Ohle, "Smoke Abatement: A Report on Recent Investigations Made at Washington University" 82

Part 5

CONSERVATION, PRESERVATION, AND HETCH HETCHY 85

Warren Olney, "Water Supply for the Cities About the Bay of San Francisco" 87

E. T. Parsons, "Proposed Destruction of Hetch-Hetchy" 91

John Muir, "Hetch Hetchy Valley" 94

Bibliographical Essay 103

Index 107

FOREWORD

Revisiting Origins: Questions That Won't Go Away

WILLIAM CRONON

WITH THE PUBLICATION of David Stradling's *Conservation in the Progressive Era: Classic Texts,* I'm proud to announce that we're adding a new kind of book to our series of Weyerhaeuser Environmental Classics. Our goal in these short volumes will be to revisit pivotal moments in environmental history so as to make it possible for readers to remind themselves of past ideas, events, and controversies that continue to shape our world today. By reprinting in an inexpensive format original documents that are the essential raw materials for experiencing history at first hand, we hope to provide a valuable resource both for students and for general readers eager to gain a richer and more nuanced sense of the environmental past. Like all the other books in the Weyerhaeuser series, the classic texts we're reprinting have been selected not just for their intellectual significance, but for their liveliness and their ability to provoke unexpected ideas and questions. By juxtaposing the familiar with the unfamiliar, the predictable with the unpredictable, we hope these anthologies will enable readers to pursue for themselves the insights and interpretations that are the heart of all historical inquiry.

It would be hard to imagine a better way to launch these new collections of classic texts than with a volume on conservation during the Progressive Era. We're lucky to have recruited David Stradling as the editor for this anthology, since he is both a master teacher and the author of the book *Smokestacks and Progressives: Environmentalists, Engineers, and Air Quality in America, 1881–1951* (Johns Hopkins University Press, 1999), which has significantly broadened scholarly understanding of Progressive conservation. In the pages that follow, he offers precisely what we hope to see in all of these anthologies: brief editorial overviews of the key issues, along with commentaries on individual documents; well-known primary sources without which no understanding of the subject would be complete; and almost forgotten documents, unfamiliar even to knowledgeable readers, designed to reach beyond conventional interpretations.

Why do we still care about Progressive conservation more than a century after its birth? At one level, the answer is simple: it was the first large-scale national political movement in American history that sought to grapple with environmental dilemmas like waste, pollution, resource exhaustion, and sustainability which continue to bedevil us today. We still invoke figures like Gifford Pinchot and John Muir—and, standing between them, Theodore Roosevelt—as the symbolic icons of debates about conservation and preservation that have never been finally resolved. One controversy in particular, about whether the city of San Francisco should be permitted to build a dam in Hetch Hetchy Valley within the boundaries of Yosemite National Park, has attained almost mythic status as a classic dispute between competing visions of how human beings should or should not use and modify the natural world.

No survey of American environmental history could possibly ignore these familiar stories. Because Stradling recognizes this, he gives us essential passages from Roosevelt, Pinchot, and Muir. He provides exemplary documents from both sides of the angry debates over Hetch Hetchy. He explains why Progressive conservation emerged when it did, and why so many Americans became concerned about natural resources at this particular moment. All the traditional themes are here.

But Stradling is not content merely to serve up familiar stories and famous men. He and other environmental historians have been critiquing this traditional picture of Progressive conservation for nearly a generation, and the documents in this collection demonstrate why. Although older historiography has often located the environmental controversies of this era primarily in rural and wilderness landscapes—amid forests, ranches, mines, dammed rivers, and national parks—Stradling demonstrates that cities and suburbs also became battlegrounds for new ideas about pollution and resource use. Although most histories of conservation give pride of place to elite male figures like Pinchot and Muir, Stradling reminds us that women played important roles too, as did members of the working class. Some of the most innovative environmental historical scholarship of the recent past, associated with the work of Louis Warren, Karl Jacoby, Mark Spence, Richard Judd, and others, has identified the ways in which working-class rural communities became the victims, in their own eyes at least, of professional-class notions of conservation emanating from far-away urban areas. Voices from these communities are also represented in this collection.

Read together, the net effect of these documents is to suggest just how complicated the early conservation movement truly was, and just how important its questions and dilemmas remain for us today. Those who began to call themselves "environmentalists" in the decades following the Second World War built on a long tradition of earlier activism that reached well beyond the conservation of natural resources. Concerns about pollution and human health are much older than we sometimes imagine, and the impact of environmental change on disempowered communities became a cause for concern long before the movement called "environmental justice" emerged during the late twentieth century.

Perhaps most intriguingly, these documents also serve as a reminder that early environmental activists struggled much as we do with the challenge of protecting nature for its own sake while also using nature to meet human needs. Today, we often see conservation and preservation as being starkly opposed to one another, with conservationists arguing for the efficient use of natural resources and preservationists arguing for the protection of natural systems from just such use. This is not new. Both impulses were fully articulated in the early political debates about environmental protection that are so well represented in these pages.

Because some of the most heated controversies of the late twentieth and early twenty-first centuries have seemingly pitted advocates of old-style conservation against advocates of new-style environmentalism, it's all too easy to imagine that the two have always been in stark opposition to one another. In fact, they spring from common roots, and often embody remarkably similar values. The larger-than-life drama of Hetch Hetchy encourages us to view Gifford Pinchot and John Muir as adversaries diametrically opposed to each other's points of view, tempting us to forget that they were once strong allies, and not by accident. Use and non-use are both essential, and a key goal of environmentalism is presumably to make sure that both are done wisely. An obvious lesson is that both conservation and preservation remain central to any comprehensive vision of environmental protection, as does the challenge of serving social justice at the same time. Among the greatest virtues of the classic texts that David Stradling has assembled on these pages is their reminder that we are not the first generation to have grappled with these difficult issues, and that we can still learn much from those who have gone before.

PREFACE

THE CONSERVATION MOVEMENT left a wealth of documentary evidence, dating from George Perkins Marsh's 1864 *Man and Nature* through at least 1972 and the Club of Rome's publication of *The Limits to Growth*. Between these dates an American conservation movement grew and at times flourished. I have purposefully chosen documents from a rather narrow chronological period, however, with most of the selections authored between 1904 and 1912, during the height of the conservation movement's national influence. I hope the collection's narrow chronological focus will encourage students to question how environmental activists acquired so much political influence during this critical period, and, perhaps, how successful environmental reform encouraged other social and political changes. In addition, I have selected documents largely from published sources, many of them widely circulated and read during the Progressive Era. I have purposefully chosen writings that were part of the political discourse to focus on conservation as a political process. I hope this collection draws the reader's attention to effective political rhetoric: the use of language to mold public opinion and effect change.

Although I have kept the range of documents narrow in these important ways, I have also attempted to broaden understanding of the conservation movement through the inclusion of differing perspectives on the movement, including those left out of many traditional narratives, such as those harmed by conservation policies. I have included documents from a variety of Progressive Era environmental reform efforts, including the anti-smoke movement, to reveal how important conservationism had become to urban environmental movements in this era. With these goals in mind, then, the collection contains expected voices, including President Theodore Roosevelt, Chief Forester Gifford Pinchot, and the Sierra Club's John Muir, but also some unexpected voices, including antismoke activist Charles Reed, Progressive reformer Mary Ritter Beard, and conservation opponents such as H.J.M. Mattes of Fort Collins, Colorado. Despite my efforts at inclusiveness, some voices remain silent here, including Native Americans, many of whom were negatively affected by conservation policies, but none of whom, to my knowledge, left an extensive written critique of conservation in this era.

All of the writings included here are excerpted from larger documents. In abbreviating these pieces I have attempted to intervene as little as possible, generally retaining paragraphs in their entirety, for example, so that readers can gain a more complete sense of the documents, both their arguments and style. Partly as a consequence of my attempt to maintain the documents' integrity, many of the readings contain what readers will rightly interpret as unnecessary detail, to which I have added few explanatory footnotes. This too is part of my philosophy of limited editorial intervention, an attempt to keep myself as much removed from the documents as possible. In a few instances, however, I have corrected obvious typos contained in the originals.

I am greatly indebted to several individuals for their help in creating this reader. First, Bill Cronon suggested this project to me and suggested some of its contents, including two of the Hetch Hetchy readings. As series editor he has guided the project from the beginning; it simply would not exist without his aid and encouragement. Second, Karl Jacoby gave me so much help I am rather embarrassed that he is not listed as co-editor. Not only did Karl suggest three of the documents included here (all three of which I would not have located myself), but he also gave me detailed comments on every aspect of this work. Karl's expertise in conservation complemented my own and together we developed a more inclusive book. I am also indebted to Nancy H. DeStefanis and Dan Philippon for helping me locate Mabel Osgood Wright's writings. I am grateful to Julidta Tarver at the University of Washington Press for shepherding this project through the entire publication process and for her helpful criticisms and unfailing encouragement. I am indebted to Kerrie Maynes for her careful and thorough editing of the manuscript. Finally, I am most indebted to Jodie Zultowsky, who aided me in this project as she does in all things.

Conservation
in the Progressive Era

Introduction

ONLY TWICE IN American history have environmental issues worked their way toward the top of the national political agenda: first, in the Progressive Era, roughly the two decades surrounding 1900, during what we call the conservation movement; and, second, in the 1960s and 1970s, during the florescence of the modern environmental movement. While these two movements share much in common, including a continuity of conservationist philosophies, this book contains documents concerning only the former moment in time, when the conservation movement first thrust environmental issues onto the national stage and played a significant role in the creation of a broad reform era.

THE DEVELOPING CONSERVATION MOVEMENT

Conservation grew only slowly into a potent national movement, as more and more nineteenth-century Americans came to recognize the potency of a number of different environmental threats, especially the diminishment of natural resources such as timber and game. Not surprisingly, efforts to conserve natural resources long predated the nationwide movement described here, as earlier conservation attempts were local in nature. The people most familiar with the resources in question, including timber, game, and fisheries, worked toward solutions that might protect them, including restricted seasons for fish and game.[1]

Only slowly did concern for natural resources move beyond the local, with significant change coming after the 1864 publication of one key text, George Perkins Marsh's *Man and Nature.* Marsh, a Vermont politician and intellectual, had recently toured Europe, studied its history, and connected the fate of civilizations with the quality of their environments. Subtitled *Phys-*

ical Geography as Modified by Human Action, his work asserted that humanity could alter the environment in permanently damaging ways, even to the point of jeopardizing the future of environmentally destructive nations. Civilizations declined, he proposed, as they destroyed their environments, particularly through deforestation. Marsh's initial work, and his subsequent revisions, served as warnings to Americans that profligate waste would threaten the nation's future prosperity.

Marsh's ideas were instantly influential, particularly among people already concerned with forests. After reading *Man and Nature,* Franklin B. Hough, of the American Association for the Advancement of Science, called upon Congress to form a national forestry commission in 1873. When Congress did so three years later, Hough became the federal government's first forestry agent, and later its first chief forester. In this position, Hough began the task of collecting information on the nation's forests and publicizing the rapid pace of their consumption, particularly through four volumes of *Report Upon Forestry,* all published before 1885.

Deforestation posed particular problems for water supplies, as George Perkins Marsh had warned, and his ideas found resonance in the movement to protect forested watersheds. The earliest such effort came in New York State, where concerned citizens successfully protected the Adirondack Mountains, first in a state forest preserve in 1885, and later in a state park, to be kept "forever wild" as required by the state constitution in 1894. Adirondack protection came largely out of the concern for the water that flowed out of the mountains, particularly the Hudson River and other tributaries to the economically critical Erie Canal system.

In the late 1800s, more Americans began to express concerns regarding several environmental issues, some of which Marsh had clearly articulated. But with such a vast landscape, so much sparsely settled land, most Americans assumed the nation's natural resources would be endlessly abundant. Only remarkable shocks could begin to convince Americans otherwise. One such shock came in 1890, when the Census Bureau announced that the American frontier had come to a close—no longer could the nation point to a line on the map beyond which open lands beckoned adventurers, speculators, and hard-working pioneers. The far-off point in time when the nation would reach its limits of growth suddenly seemed all too near at hand.

Other shocks came through local experiences, especially the diminishment of regional resources. Indeed, the disappearance of forests in the East and upper Midwest heightened concern about the pace of the nation's resource consumption. Clearcuts left devastated landscapes, leading to erosion, flooding, and even massive forest fires, as dry debris left by careless timbermen fueled remarkable blazes. Although these fires could plague any cutover land, the deadliest blaze occurred in Peshtigo, Wisconsin, in 1871, where up to 1,500 people perished. Obviously such fires served as shocks as well, indicating that common forestry practices had failed.

With the influence of Marsh and growing knowledge of European forestry, observers of the rapid consumption of great American forests could both fear long-term consequences and suggest alternatives. Indeed, European foresters, working under conditions of scarcity unheard of in the United States, had begun to develop techniques that could greatly improve the efficiency of timber harvests. Many of the early American foresters, including German-born Bernard Fernow, who headed the United States Division of Forestry in 1879, received training in Europe, learning how properly managed forests could remain permanently productive.

With increasing awareness of the rapidity of forest consumption and the terribly wasteful practices that helped speed the destruction, the federal government acted to protect forests still in the public domain. In 1891, Congress passed the Forest Reserve Act, allowing presidents to set aside federal lands, and shortly thereafter Benjamin Harrison became the first to do so. In 1897, the Forest Management Act actually gave the Department of the Interior a mission regarding these lands, empowering it to regulate grazing, timber harvests, and other natural resource extraction in the reserves.

Rapid settlement in the West during the late 1800s led to other environmental concerns as well, including the need for better water management. Scarce rainfall impeded agricultural growth in much of the West, and to encourage a fuller occupation of these arid regions many Westerners called for a national effort to capture available waters, impound them in reservoirs, and distribute them to new irrigated farms. In 1902, Nevada Representative Francis Newlands won national intervention through his support of the Newlands Reclamation Act, which allowed the federal government to begin planning and building dams and canals to expand agriculture in the arid West. While most of the building actually took place decades later, the Newlands Act laid the legal groundwork for the transformation of the West through the conservation of water in arid environments, and it held out the promise of the further expansion of agriculture in the West.

The nation's remarkably rapid nineteenth-century economic and demographic expansion also created concern for the preservation of wild and scenic places. Thinking beyond the consumption of natural resources, many Americans began to concern themselves with the fact that an ever-growing nation would eventually consume its last open expanses. The artist George Catlin had expressed an interest in preserving the American wilderness as early as 1832, and by the late 1800s presidents began to do so in the world's first national parks. Beginning with the designation of Yosemite Valley as parkland in 1864, and more importantly with President Ulysses S. Grant's creation of the expansive Yellowstone National Park in 1872, a movement to set aside spectacular landscapes eventually led to the creation of dozens of national parks and national monuments by the 1920s, the majority of them in the mountains of western states.

If concern for the preservation of particular landscapes grew, so too did concern for individual species, especially those under intense pressure from hunting and habitat destruction. The dramatic decline of the Great Plains' once-vast bison herds to the point of near extinction in the 1880s served as a warning that any game species might be perilously overhunted. Perhaps even more dramatic, passenger pigeon flocks, which had once numbered into the billions of birds, disappeared from the eastern United States and passed into extinction when the last member of the species died in the Cincinnati Zoo in 1914. With such remarkable declines as evidence, many individuals, particularly hunters, took up the effort to protect favored game species, especially deer and waterfowl. Activists, working effectively through sportsmen's clubs and within the sporting press (including George Bird Grinnell's *Forest and Stream,* begun in 1873) and through the Audubon Society (created first in 1886 and then re-created in 1905) forced the creation of hunting regulations at the state and local levels and encouraged preservation measures at the national level.

While late-nineteenth-century conservation concerns clearly focused on the West, where changes due to growth were particularly evident, those living in the East also found reason for worry. Dwindling forests, diminishing game species, and threatened water supplies sparked concern throughout the nation. In addition, citizens living in the nation's industrial cities, particularly in the East and Midwest, began to express concern for the urban environment. Especially in the 1890s, middle-class urbanites began to demand effective regulation of smoke emissions, complaining about foul air and dirty soot. By 1900, many large cities had passed new antismoke ordinances, and the movement to clear the air would soon gain national momentum. Just as important, and perhaps more effective, a movement to clean city streets had led to numerous local reforms in garbage collection, street sweeping, and sewage disposal, largely in the name of protecting public health. Intensifying industrialization and urbanization had created many urban environmental problems, and conservation rhetoric would play a critical role in the efforts to abate them.

Altogether, then, the pressures of industrialization and rapid consumption, felt both in the city and the country, posed serious and growing environmental threats in the late 1800s. While the nation contained ample timber, water, coal, and other vital resources for the near term, more and more Americans began to fear national environmental degradation. Many feared that wasteful consumption threatened premature restrictions on the nation's growth and development.

By the time Theodore Roosevelt entered office in 1901, then, the nation had experienced a long, if uneven, growth in public awareness of environmental problems, and during Roosevelt's administration the conservation movement flourished, as concern for disparate environmental problems came together to energize a remarkably successful reform effort. Roosevelt's interest in conservation derived not only out of a fear that the nation would soon deplete its natural resources, but also from less tangible concerns about America's future. Raised in a wealthy New York family, Roosevelt's privileged birth gave him an opportunity to explore the natural riches of the West as a tourist. He feared, however, that most urban Americans would lose touch with their nation's rugged past, and that the long connection with the frontier and its gifts of self-reliance, masculinity, and democracy would be severed. To prevent cultural decline, Roosevelt envisioned a new "strenuous life," calling for vigorous living, particularly in connection with nature. For Roosevelt, then, conservation would mean the unending presence of a frontier preserved by the federal government.[2]

In addition to Roosevelt himself, the conservation movement gained another influential voice in his administration, that of Chief Forester Gifford Pinchot, who had gained that post when Bernard Fernow resigned in 1898. Like Roosevelt, Pinchot had been born into a wealthy East Coast family, and he lived a privileged life of travel and education. After studying forestry in Europe and gaining practical experience in the United States, Pinchot exerted remarkable influence over American conservation policies, which he claimed were primarily about development—the proper management of resources to ensure efficient use. Indeed, for many people, the words of Pinchot offer the clearest definition of the movement. In *The Fight for Conservation,* excerpted at length below, Pinchot stated succinctly: "Conservation means the greatest good to the greatest number for the longest time."[3]

As forest historian Michael Williams notes, Pinchot "claimed that he initiated conservation," but he actually drew on a body of forestry expertise that dated back to the 1870s, when Fernow, Charles Sargent, and other foresters began to apply European forestry techniques to the United States and lobby for governmental protection of remaining forests.[4] Much of Pinchot's success, then, came from his political skills in selling conservation and himself. This success reminds us that in the end conservation was a political movement as much as a scientific one, and as such it required effective use of political rhetoric. Together, Pinchot and Roosevelt popularized conservation rhetoric—much of it based in science—and developed effective environmental policies.

Conservation under Roosevelt witnessed a series of victories. Although previous and subsequent presidents would set aside national forests too, Roosevelt was particularly vigorous in protecting lands within the public domain. By 1916, Roosevelt and four other presidents had set aside 176 million acres of western forests.[5] In 1903, Roosevelt also set aside the nation's first wildlife refuge, Pelican Island, Florida, the first of fifty-three reserves he created to protect habitat for wildlife, especially in key breeding grounds. An avid hunter and a founding member of the conservation-minded Boone and Crockett Club with his friend George Bird Grinnell, Roosevelt supported game protection, both through the creation of wildlife preserves and the enforcement of the Lacey Act of 1900, which outlawed interstate shipments of protected birds. The Antiquities Act of 1906 allowed Roosevelt to create national monuments on federal lands, protecting "historic landmarks, historic and prehistoric structures, and other objects of historic or scientific interest." Roosevelt used this act to good effect, creating eighteen national monuments, some of which later became national parks, including the Grand Canyon, which Roosevelt protected in 1908. Taken together, the actions of Roosevelt's administration represented a high-water mark in environmental activism, setting aside lands for protection and establishing more effective resource regulations.

As effective politicians, Pinchot and Roosevelt both made good use of what Roosevelt called the "Bully pulpit," frequently calling the public's attention to conservation issues. In 1908, Pinchot organized a Governors' Conference on the Conservation of Natural Resources, held at the White House.

One month later, Roosevelt appointed a National Conservation Commission headed by Pinchot and staffed by congressmen and federal bureaucrats. The following year that commission published the first national inventory of natural resources in three volumes, parts of which are excerpted below. The Governors' Conference and the commission reports helped ensure that conservation would be one of the most salient topics in Roosevelt's last year in office, and that his successor, William Howard Taft, would continue to press conservation policies.

Despite all of this governmental activism, the conservation movement was not simply a timely and reasonable reaction to resource depletion. Instead, it grew in conjunction with a broader reform movement called progressivism. Although historians still debate what exactly Progressive reform was, it had some obvious themes. Developing in the 1890s, progressivism expressed an interest in improving society through a series of reforms, most of which expanded the power of government at all levels—federal, state and local—to solve the many problems associated with industrial and urban growth. By the 1910s, any number of people could claim to be Progressives, and the diversity of activists at work speaks to the strength of the reform fervor. Among prominent reformers was Jane Addams, founder of Chicago's Hull House in 1889, and the leading voice in the settlement house movement, which sought to bring middle-class values to urban immigrant neighborhoods. Other reformers included muckraking journalists such as Ida Tarbell, who sought reforms by exposing corrupt practices among large corporations, and, of course, politicians such as Roosevelt, who saw a need for activist government. While a real diversity of Progressives existed, most were wealthy, or at least middle class, and urban.

The reforms supported by Progressives were diverse as well. Some sought to regulate powerful corporate monopolies and trusts, made more successful under the Clayton Antitrust Act of 1914; others sought to improve the lives of the urban poor, particularly through neighborhood improvements, including new schools, playgrounds, and libraries; and others sought to topple wasteful and criminal political machines that controlled so many American cities, especially through democratic reforms such as secret ballots. Other reforms brought new rules for child labor and shortened workweeks, and the Pure Food and Drug Act of 1906 protected consumers from dangerous or fraudulent products. In addition, Progressives concerned themselves with a number of environmental issues, including conservation, which became one of the hallmarks of the larger reform movement.[6]

Although it is difficult to generalize about these diverse reforms, progressivism was a political response to the pressures industrialization had exerted on the nation, including a growing sense of chaotic, wasteful change. Since both conservation and the broader progressivism strove to bring order out of chaos, they shared a language of reform. At the heart of this language lay two concepts: democracy and progress. From these spun a host of related concerns: antimonopoly, the people versus "the interests," efficiency, purity, and the need for scientific understanding, often gained through survey or inventory. The era also witnessed a remarkably strong faith in government (as representative and protector of the people) and science (as the producer of knowledge and solutions). At the core of the national discourse on democracy and progress lay another key term—civilization. Many middle-class Americans had come to believe that their nation was rapidly approaching a new apex in civilized society, and they consciously and unabashedly worked toward that goal. Conservation—improved efficiency and order—would strengthen America's claim to civilization's crown.

Supporters of conservation helped create the reform lexicon, and Progressive Era activists used much of it in support of various reforms. Activists sprinkled their language with praise for "the people," "efficiency," "civilization," and fear of their opposites, "the interests," "disorder," and "savagery." Note how frequently and flexibly this language of democracy and progress appears in the following documents. Supporters of reforms as different as regulating corporate monopolies, protecting forests, and building better schools could speak about their efforts as defending the rights of the people and contributing to the rise of the American civilization. A strong argument can be made that the success of the conservation movement, and Progressive reformism generally, had much to do with the portability of this reform language and the popularity of the philosophical concepts behind it.[7]

By no means did conservation disappear after Roosevelt's administration ended in 1909, as subsequent presidents continued to implement conservation policies, including setting aside more national forests and creating more national parks. Indeed, conservationism flourished again under Franklin Roosevelt's administration in the 1930s, and conservation rhetoric gained a national audience as the country developed a stronger, broader environmental movement in the 1960s. In that decade, the utilitarianism of the conservation movement became one of a number of distinct philosophical threads that came together to create the modern environmental movement. Another thread was spun from Romanticism, with roots in nine-

teenth-century Europe, which continued to encourage aesthetic apprecia-
tion of nature—particularly wilderness. In addition, concerns for human
health had long encouraged environmentalist concerns about pollution, and
by the 1960s this particular thread of environmentalism had gained promi-
nence due to the development of new toxic threats, including nuclear fall-
out and persistent pesticides. Other threads of environmentalism included
the moralistic animal rights movement and, importantly, a scientific, eco-
logical approach to understanding environmental threats, both of which
gained broader support in the 1960s. Together with the persisting conservationist ideal of efficiency, these threads helped create the multifaceted environmentalist movement, resulting in a flurry of significant legislation,
including the Wilderness Act (1964), the Clean Air (1970) and Clean Water
Acts (1972), and the Endangered Species Act (1973).

INTERPRETING CONSERVATION

Given its political nature, it is not surprising that conservation has been subject to a number of different interpretations, beginning with those engaged
in the movement itself. Even as the movement gathered momentum, par-
ticipants attempted to define conservation favorably, including Pinchot in
The Fight for Conservation (1910) and University of Wisconsin President
Charles Van Hise in *The Conservation of Natural Resources in the United
States*, published in the same year. However, Pinchot's later work, a lengthy
autobiography and examination of the conservation movement, *Breaking
New Ground*, published in 1947, shortly after his death in the fall of the previous year, provided the clearest articulation of conservation as a utilitarian, democratic movement. Not surprisingly, Pinchot and Roosevelt were
at the center of this narrative history, their careers clearly marking the ebb
and flow of the movement. Also, in this conception of the movement conservation was fundamentally democratic, designed to protect the nation's
resources from potential despoilers—particularly large monopolies that
threatened to race through the nation's forests and occupy the West's best
dam sites. Certainly a democratic philosophy underlay conservationism; note
Pinchot's own phrase, "for the greatest number."

The first historian to offer a full critique of this early understanding of
conservation was Samuel Hays, whose *Conservation and the Gospel of Efficiency* has remained the most important work concerning the conservation
movement since its publication in 1959.[8] Hays powerfully argued that the

conservation movement was part of a larger societal shift toward scientific expertise and organization. Conservationism was a positive philosophy designed to ensure continued production through ever-increasing efficiency. In Hays's telling, experts, particularly engineers and foresters, were the heroes of the conservation movement, applying science to natural resource exploitation, bringing order and permanence to consumption. Engineers designed and built dams and irrigation systems to order the use of otherwise "wasted" water resources. Other experts, biologists and foresters particularly, managed game and timber resources, setting restrictions and guidelines for their harvest. Still other experts designed more efficient technology in an effort to conserve coal and other production resources. In sum, according to Hays conservation was an expert-driven effort to sustain industrial consumption, not the rights of "the people." That Hays's work is still the most complete interpretation of Progressive Era conservation suggests how well he articulated some fundamental truths about the movement.

More recently historians have begun to critique both the narrowness of Hays' interpretation and the narrowness of the conservation movement itself. A diversity of opinions always existed as to what conservation was or should be. In addition, and perhaps more important, people debated whether or not conservationism inspired necessary or effective policy. Despite the hopeful, democratic tone evident in many of the documents included here, and the writings of other Progressive conservationists, we cannot forget their limited visions. Recent scholarship has further attacked the democratic notions of conservationists, noting particularly their ethnocentricity. Conservationists such as Roosevelt and Pinchot offered an overwhelmingly eastern, wealthy, and white perspective on environmental issues. Regardless of the efficacy of conservation policies in protecting the environment, they generally did not sufficiently protect the rights of residents in the West, particularly Native Americans, who could rightly see conservation legislation as simply another aspect of an ongoing imperialist invasion. In addition, rural whites could see conservation policies as unnecessary federal (or state) meddling in local affairs. We must understand that conservation policies did bring tangible losses to more than just large, faceless corporations bent on extracting profits from resource exploitation. The rural poor, and particularly Native Americans, lost access to traditional resources.

The conservation movement was clearly a component of the arriving national market, a regulatory facet in the quickening capitalist exploitation of natural resources, and as such it created winners, losers, and conflict.[9]

In addition, conservation was more than just a national effort to protect western natural resources. By the 1910s, the multifaceted movement included efforts to conserve human resources, beauty, and "national efficiency." Eventually, conservationist language pervaded many of the environmental reform movements gathering strength in American cities, including efforts to develop better city planning, improve public health, control smoke, and make cities generally more beautiful and efficient.

DEFINING CONSERVATION

Since conservationism did not disappear after the 1910s, and indeed changed over time, considerable confusion persists as to what we ought to consider "conservation." Even historians have used the word to mean very different things. The *Dictionary of American History,* for example, defines conservation as "the wise use of resources," but then quickly points out that one group of conservationists has "promoted preservation or nonuse."[10] In other words, the term does double (and conflicting) duty—signifying both a movement to promote efficient use and the preservation movement that struggled against that use. Some historians have attempted to ease the confusion by casting conservation and preservation as distinct threads within the broader environmental movement. And, clearly, the two philosophies are not in constant conflict—preservation can be a technique used to support specific conservation goals. For example, habitat preservation serves the interest of wildlife conservation.

Adding to the confusion, history textbooks tend to lump together the two major efforts regarding the protection of federal lands—conservation in national forests and preservation in national parks—largely because the two efforts developed at about the same time. Roosevelt, like most Progressives, supported both conservation and preservation, for they both served a purpose in controlling change in a rapidly growing nation, and so in the Progressive Era preservationism flourished along with conservationism.

Although preservationism and conservationism did not always conflict, we should be careful to avoid lumping all environmental philosophies together. John Muir's voice is found in this book of readings, in a well-known passage concerning the preservation of Hetch Hetchy Valley in Yosemite National Park, but not because he represents a preservationist "wing" of conservation. Rather, in this instance his is a voice against conservation—against the storage of a natural resource within a national park. As a founder of the

Sierra Club, and its guiding force even after his death in 1914, Muir surely supported conservation in some instances, but here, and most famously, he spoke eloquently against its philosophy and goals. Still, note how frequently the documents included here use the two words interchangeably or in support of one another, suggesting that we ought not consider these two efforts as the central conflict within the evolving environmental movement.

QUESTIONS FOR THE DOCUMENTS

Within the following documents are hints on how one might answer the most basic questions concerning conservation. What was the conservation movement? As these documents make clear, even during the relatively short Progressive Era, people used the term "conservation" variously. Pinchot and Roosevelt were hardly the only ones to harness the power of this term to further their own goals. Many others clearly used the term and the ideas behind it because it had already gained influence, suggesting the success of conservation in this era. Why was the conservation movement so successful? Given the outline of events listed above, clearly the government took many positive steps to protect natural resources in this era: the passage of environmental legislation and executive action would go unmatched until the 1970s. Conversely, the movement had some glaring failures, particularly in securing resources for democratic distribution and in improving urban air quality. Why did the movement fail in these key areas?

While historians tend to judge the conservation movement a success, at least as far as it went, many at the time found only disaster in the results. According to some observers, the conservation movement created potent threats to democracy, both in the short term, as new laws kept certain people away from their traditional resources, and in the long term, as some feared a growing, distant bureaucracy. So, we might also ask, what can we learn from the opponents of conservation policies?

Our own environmental sensibilities may encourage us to misread the following documents, to think them conservative or naïve. While it makes sense to connect Progressive Era conservation with previous and subsequent environmental philosophies, to understand the movement we must place it squarely in its own time. The authors of the following documents held different assumptions than readers do today about what problems truly threatened their nation and what types of solutions might actually work. Subsequent generations have learned much from the conservation movement, but perhaps not as much as they might. As Americans continue to work to protect their environment, perhaps the most important lesson concerns effectiveness. How were conservationists in the Progressive Era able to connect their interests to so many other interests, creating broad support for environmental protection?

NOTES

1. See particularly Richard W. Judd, *Common Lands, Common People: The Origins of Conservation in Northern New England* (Cambridge, Mass.: Harvard University Press, 1997).

2. See Gail Bederman, *Manliness and Civilization: A Cultural History of Gender and Race in the United States, 1880–1917* (Chicago: University of Chicago Press, 1995), 192–96.

3. Gifford Pinchot, *The Fight for Conservation* (Seattle: University of Washington Press, 1967), 48. (Originally published 1910.)

4. Michael Williams, *Americans and Their Forests: A Historical Geography* (New York: Cambridge University Press, 1989), 402.

5. John Opie, *Nature's Nation: An Environmental History of the United States* (New York: Harcourt Brace College Publishers, 1998), 379.

6. On progressivism, see particularly Robert Wiebe, *The Search for Order, 1877–1920* (New York: Hill & Wang, 1967) and Daniel T. Rodgers, *Atlantic Crossings: Social Politics in a Progressive Age* (Cambridge, Mass.: Belknap Press of Harvard University Press, 1998).

7. On the importance of language in politics, see Daniel T. Rodgers, *Contested Truths: Keywords in American Politics Since Independence* (New York: Basic Books, 1987).

8. Samuel Hays, *Conservation and the Gospel of Efficiency* (Cambridge, Mass.: Harvard University Press, 1959).

9. See Louis S. Warren, *The Hunter's Game: Poachers and Conservationists in Twentieth-Century America* (New Haven, Conn.: Yale University Press, 1997) and Karl Jacoby, *Crimes Against Nature: Squatters, Poachers, Thieves, and the Hidden History of American Conservation* (Berkeley: University of California Press, 2001).

10. George Carney, "Conservation," *Dictionary of American History* (New York: Charles Scribner's Sons, 1976), 187.

Part 1

DEFINING AND DEBATING

CONSERVATION

The following documents reveal the contested nature of conservation as it gained popularity during the early 1900s. The first four documents present the long-standing, dominant definition of conservation from different perspectives, all of which praise the expansion of governmental power. First, former Chief Forester Gifford Pinchot offers the most repeated definition, linking conservation to development and democracy. Importantly, Pinchot also writes of conservation as a broad movement, one that could improve "national efficiency." This was a political document published in 1910, shortly after Pinchot's ouster from the Forest Service and as Theodore Roosevelt contemplated another run for office.

President Roosevelt authored the second document, a "Special Message" written to introduce the *Report of the National Conservation Commission* in 1909. This three-volume report described the extent of the nation's natural resources, including minerals, agricultural lands, forests, and fresh water. The report also indicated in what areas the nation needed conservation measures. In addition to articulating many of the themes found in Pinchot's writings, Roosevelt's introduction emphasized the importance of Government (note the capital G), laying the philosophical groundwork for an expanded and empowered bureaucracy acting to protect the "industrial liberty" of the nation's citizens. The third piece comes from a leading irrigation crusader, William Smythe, publisher of *The Irrigation Age*, whose book-length treatment of the issue, *The Conquest of Arid America,* became one of the clearest calls for governmental involvement in reclamation projects in the American West. In this excerpt, Smythe emphasizes the democratic nature of irrigation projects.

In the fourth document, the *Ladies' Home Journal* explores the relationship between women and conservation, but in a very condescending tone.

Ignoring the important political contributions women had already made to the movement, the column "What is Meant by Conservation?" instead focuses on the numerous ways women practiced conservation in their everyday lives as homemakers. According to this definition, Conservation, with a capital *C,* is nothing more than economy practiced by Government, with a capital *G.*

The final two documents reveal something of conservation's opponents. In an article from the *North American Review,* the modestly successful eastern author George Knapp condemns the rhetoric of conservationists, claiming that the nation's natural resources had not been wasted, only used. The last document, a letter from H.J.M. Mattes, of Fort Collins, Colorado, to the editor of Denver's *Rocky Mountain News,* protests federal conservation efforts in the West. The pro-conservation journal *Forestry and Irrigation* republished this letter as evidence of the "humorous" opposition conservation faced in the mountain region. Mattes, of course, found nothing funny about federal intervention in Colorado's natural resource use. Both Mattes and Knapp reveal how some citizens could feel a greater threat from the expansion of governmental power than from the depletion of natural resources, even to the point of suggesting conservation threatened the nation's founding principles.

These documents, read together, describe conservation as a movement emphasizing development, positive governmental activism, and the protection of the commonweal, perhaps at the expense of the individual. In addition, Roosevelt himself rests at the center of this definition of conservation. Both in and out of office, Roosevelt and Pinchot worked hard to define conservation as *their* policy, to proclaim the success of that policy, and to suggest that other policies threatened the nation's survival. Opponents, of course, were forced to battle both the popularity of Roosevelt and his policies.

Principles of Conservation

GIFFORD PINCHOT

THE PRINCIPLES WHICH the word Conservation has come to embody are not many, and they are exceedingly simple. I have had occasion to say a good many times that no other great movement has ever achieved such progress in so short a time, or made itself felt in so many directions with such vigor and effectiveness, as the movement for the conservation of natural resources.

Forestry made good its position in the United States before the conservation movement was born. As a forester I am glad to believe that conservation began with forestry, and that the principles which govern the Forest Service in particular and forestry in general are also the ideas that control conservation.

The first idea of real foresight in connection with natural resources arose in connection with the forest. From it sprang the movement which gathered impetus until it culminated in the great Convention of Governors at Washington in May, 1908. Then came the second official meeting of the National Conservation movement, December, 1908, in Washington. Afterward came the various gatherings of citizens in convention, come together to express their judgment on what ought to be done, and to contribute, as only such meetings can, to the formation of effective public opinion.

The movement so begun and so prosecuted has gathered immense swing and impetus. In 1907 few knew what Conservation meant. Now it has become a household word. While at first Conservation was supposed to apply only to forests, we see now that its sweep extends even beyond the natural resources.

The principles which govern the conservation movement, like all great and effective things, are simple and easily understood. Yet it is often hard

From *The Fight for Conservation* (Seattle: University of Washington Press, 1967), 40–52. (Originally published 1910.)

to make the simple, easy, and direct facts about a movement of this kind known to the people generally.

The first great fact about conservation is that it stands for development. There has been a fundamental misconception that conservation means nothing but the husbanding of resources for future generations. There could be no more serious mistake. Conservation does mean provision for the future, but it means also and first of all the recognition of the right of the present generation to the fullest necessary use of all the resources with which this country is so abundantly blessed. Conservation demands the welfare of this generation first, and afterward the welfare of the generations to follow.

The first principle of conservation is development, the use of the natural resources now existing on this continent for the benefit of the people who live here now. There may be just as much waste in neglecting the development and use of certain natural resources as there is in their destruction. We have a limited supply of coal, and only a limited supply. Whether it is to last for a hundred or a hundred and fifty or a thousand years, the coal is limited in amount, unless through geological changes which we shall not live to see, there will never be any more of it than there is now. But coal is in a sense the vital essence of our civilization. If it can be preserved, if the life of the mines can be extended, if by preventing waste there can be more coal left in this country after we of this generation have made every needed use of this source of power, then we shall have deserved well of our descendants.

Conservation stands emphatically for the development and use of waterpower now, without delay. It stands for the immediate construction of navigable waterways under a broad and comprehensive plan as assistants to the railroads. More coal and more iron are required to move a ton of freight by rail than by water, three to one. In every case and in every direction the conservation movement has development for its first principle, and at the very beginning of its work. The development of our natural resources and the fullest use of them for the present generation is the first duty of this generation. So much for development.

In the second place conservation stands for the prevention of waste. There has come gradually in this country an understanding that waste is not a good thing and that the attack on waste is an industrial necessity. I recall very well indeed how, in the early days of forest fires, they were considered simply and solely as acts of God, against which any opposition was hopeless and any attempt to control them not merely hopeless but childish. It was assumed that they came in the natural order of things, as inevitably as the seasons or the rising and setting of the sun. To-day we understand that forest fires are wholly within the control of men. So we are coming in like manner to understand that the prevention of waste in all other directions is a simple matter of good business. The first duty of the human race is to control the earth it lives upon.

We are in a position more and more completely to say how far the waste and destruction of natural resources are to be allowed to go on and where they are to stop. It is curious that the effort to stop waste, like the effort to stop forest fires, has often been considered as a matter controlled wholly by economic law. I think there could be no greater mistake. Forest fires were allowed to burn long after the people had means to stop them. The idea that men were helpless in the face of them held long after the time had passed when the means of control were fully within our reach. It was the old story that "as a man thinketh, so is he"; we came to see that we could stop forest fires, and we found that the means had long been at hand. When at length we came to see that the control of logging in certain directions was profitable, we found it had long been possible. In all these matters of waste of natural resources, the education of the people to understand that they can stop the leakage comes before the actual stopping and after the means of stopping it have long been ready at our hands.

In addition to the principles of development and preservation of our resources there is a third principle. It is this: The natural resources must be developed and preserved for the benefit of the many, and not merely for the profit of a few. We are coming to understand in this country that public action for public benefit has a very much wider field to cover and a much larger part to play than was the case when there were resources enough for every one, and before certain constitutional provisions had given so tremendously strong a position to vested rights and property in general.

A few years ago President Hadley, of Yale, wrote an article which has not attracted the attention it should. The point of it was that by reason of the XIVth amendment to the Constitution, property rights in the United States occupy a stronger position than in any other country in the civilized world. It becomes then a matter of multiplied importance, since property rights once granted are so strongly entrenched, to see that they shall be so granted that the people shall get their fair share of the benefit which comes from the development of the resources which belong to us all. The time to do that is now. By so doing we shall avoid the difficulties and conflicts which will surely arise if we allow vested rights to accrue outside the possibility of governmental and popular control.

✳ The conservation idea covers a wider range than the field of natural resources alone. Conservation means the greatest good to the greatest number for the longest time. One of its great contributions is just this, that it has added to the worn and well-known phrase, "the greatest good to the greatest number," the additional words "for the longest time," thus recognizing that this nation of ours must be made to endure as the best possible home for all its people. ✳

Conservation advocates the use of foresight, prudence, thrift, and intelligence in dealing with public matters, for the same reasons and in the same way that we each use foresight, prudence, thrift, and intelligence in dealing with our own private affairs. It proclaims the right and duty of the people to act for the benefit of the people. Conservation demands the application of common-sense to the common problems for the common good.

The principles of conservation thus described—development, preservation, the common good—have a general application which is growing rapidly wider. The development of resources and the prevention of waste and loss, the protection of the public interests, by foresight, prudence, and the ordinary business and home-making virtues, all these apply to other things as well as to the natural resources. There is, in fact, no interest of the people to which the principles of conservation do not apply.

The conservation point of view is valuable in the education of our people as well as in forestry; it applies to the body politic as well as to the earth and its minerals. A municipal franchise is as properly within its sphere as a franchise for water-power. The same point of view governs in both. It applies as much to the subject of good roads as to waterways, and the training of our people in citizenship is as germane to it as the productiveness of the earth. The application of common-sense to any problem for the Nation's good will lead directly to national efficiency wherever applied. In other words, and that is the burden of the message, we are coming to see the logical and inevitable outcome that these principles, which arose in forestry and have their bloom in the conservation of natural resources, will have their fruit in the increase and promotion of national efficiency along other lines of national life.

The outgrowth of conservation, the inevitable result, is national efficiency. In the great commercial struggle between nations which is eventually to determine the welfare of all, national efficiency will be the deciding factor.

So from every point of view conservation is a good thing for the American people.

The National Forest Service, one of the chief agencies of the conservation movement, is trying to be useful to the people of this nation. The Service recognizes, and recognizes it more and more strongly all the time, that whatever it has done or is doing has just one object, and that object is the welfare of the plain American citizen. Unless the Forest Service has served the people, and is able to contribute to their welfare it has failed in its work and should be abolished. But just so far as by cooperation, by intelligence, by attention to the work laid upon it, it contributes to the welfare of our citizens, it is a good thing and should be allowed to go on with its work.

The Natural Forests are in the West. Headquarters of the Service have been established throughout the Western country, because its work cannot be done effectively and properly without the closest contact and the most hearty cooperation with the Western people. It is the duty of the Forest Service to see to it that the timber, water-powers, mines, and every other resource of the forests is used for the benefit of the people who live in the neighborhood or who may have a share in the welfare of each locality. It is equally its duty to cooperate with all our people in every section of our land to conserve a fundamental resource, without which this Nation cannot prosper.

Special Message from the President of the United States

THEODORE ROOSEVELT

I TRANSMIT HEREWITH a report of the National Conservation Commission, together with the accompanying papers. This report, which is the outgrowth of the conference of governors last May, was unanimously approved by the recent joint conference held in this city between the National Conservation Commission and governors of States, state conservation commissions, and conservation committees of great organizations

From Henry Gannett, ed., *Report of the National Conservation Commission*, vol. 1 (Washington, D.C.: U.S. Government Printing Office, 1909), 1–9.

of citizens. It is therefore in a peculiar sense representative of the whole nation and all its parts.

With the statements and conclusions of this report I heartily concur, and I commend it to the thoughtful consideration both of the Congress and of our people generally. It is one of the most fundamentally important documents ever laid before the American people. It contains the first inventory of its natural resources ever made by any nation. In condensed form it presents a statement of our available capital in material resources, which are the means of progress, and calls attention to the essential conditions upon which the perpetuity, safety, and welfare of this nation now rest and must always continue to rest. It deserves, and should have, the widest possible distribution among the people.

The facts set forth in this report constitute an imperative call to action. The situation they disclose demands that we, neglecting for a time, if need be, smaller and less vital questions, shall concentrate an effective part of our attention upon the great material foundations of national existence, progress, and prosperity.

This first inventory of natural resources prepared by the National Conservation Commission is undoubtedly but the beginning of a series which will be indispensable for dealing intelligently with what we have. It supplies as close an approximation to the actual facts as it was possible to prepare with the knowledge and time available. The progress of our knowledge of this country will continually lead to more accurate information and better use of the sources of national strength. But we can not defer action until complete accuracy in the estimates can be reached, because before that time many of our resources will be practically gone. It is not necessary that this inventory should be exact in every minute detail. It is essential that it should correctly describe the general situation; and that the present inventory does. As it stands it is an irrefutable proof that the conservation of our resources is the fundamental question before this nation, and that our first and greatest task is to set our house in order and begin to live within our means.

The first of all considerations is the permanent welfare of our people; and true moral welfare, the highest form of welfare, can not permanently exist save on a firm and lasting foundation of material well-being. In this respect our situation is far from satisfactory. After every possible allowance has been made, and when every hopeful indication has been given its full weight, the facts still give reason for grave concern. It would be unworthy of our history and our intelligence, and disastrous to our future, to shut our eyes to these facts or attempt to laugh them out of court. The people should and will rightly demand that the great fundamental questions shall be given attention by their representatives. I do not advise hasty or ill-considered action on disputed points, but I do urge, where the facts are known, where the public interest is clear, that neither indifference and inertia, nor adverse private interests, shall be allowed to stand in the way of the public good. *

The great basic facts are already well known. We know that our population is now adding about one-fifth to its numbers in ten years, and that by the middle of the present century perhaps one hundred and fifty million Americans, and by its end very many millions more, must be fed and clothed from the products of our soil. With the steady growth in population and the still more rapid increase in consumption, our people will hereafter make greater and not less demands per capita upon all the natural resources for their livelihood, comfort, and convenience. It is high time to realize that our responsibility to the coming millions is like that of parents to their children, and that in wasting our resources we are wronging our descendants.

We know now that our rivers can and should be made to serve our people effectively in transportation, but that the vast expenditures for our waterways have not resulted in maintaining, much less in promoting, inland navigation. Therefore, let us take immediate steps to ascertain the reasons and to prepare and adopt a comprehensive plan for inland-waterway navigation that will result in giving the people the benefits for which they have paid but which they have not yet received. We know now that our forests are fast disappearing, that less than one-fifth of them are being conserved, and that no good purpose can be met by failing to provide the relatively small sums needed for the protection, use, and improvement of all forests still owned by the Government, and to enact laws to check the wasteful destruction of the forests in private hands. There are differences of opinion as to many public questions; but the American people stand nearly as a unit for waterway development and for forest protection.

We know now that our mineral resources once exhausted are gone forever, and that the needless waste of them costs us hundreds of human lives and nearly $300,000,000 a year. Therefore, let us undertake without delay the investigations necessary before our people will be in position, through state action or otherwise, to put an end to this huge loss and waste, and conserve both our mineral resources and the lives of the men who take them from the earth. . . .

The function of our Government is to insure to all its citizens, now and hereafter, their rights to life, liberty, and the pursuit of happiness. If we of this generation destroy the resources from which our children would otherwise derive their livelihood, we reduce the capacity of our land to support a population, and so either degrade the standard of living or deprive the coming generations of their right to life on this continent. If we allow great industrial organizations to exercise unregulated control of the means of production and the necessaries of life, we deprive the Americans of today and of the future of industrial liberty, a right no less precious and vital than political freedom. Industrial liberty was a fruit of political liberty, and in turn has become one of its chief supports, and exactly as we stand for political democracy so we must stand for industrial democracy. . . .

We have realized that the right of every man to live his own life, provide for his family, and endeavor, according to his abilities, to secure for himself and for them a fair share of the good things of existence, should be subject to one limitation and to no other. The freedom of the individual should be limited only by the present and future rights, interests, and needs of the other individuals who make up the community. We should do all in our power to develop and protect individual liberty, individual initiative, but subject always to the need of preserving and promoting the general good. When necessary, the private right must yield, under due process of law and with proper compensation, to the welfare of the commonwealth. . . .

All this is simply good common sense. The underlying principle of conservation has been described as the application of common sense to common problems for the common good. If the description is correct, then conservation is the great fundamental basis for national efficiency. In this stage of the world's history, to be fearless, to be just, and to be efficient are the three great requirements of national life. National efficiency is the result of natural resources well handled, of freedom of opportunity for every man, and of the inherent capacity, trained ability, knowledge, and will, collectively and individually, to use that opportunity. . . .

The unchecked existence of monopoly is incompatible with equality of opportunity. The reason for the exercise of government control over great monopolies is to equalize opportunity. We are fighting against privilege. . . .

Our public-land policy has for its aim the use of the public land so that it will promote local development by the settlement of home makers; the policy we champion is to serve all the people legitimately and openly, instead of permitting the lands to be converted, illegitimately and under cover, to the private benefit of a few. Our forest policy was established so that we might use the public forests for the permanent public good, instead of merely for temporary private gain. The reclamation act, under which the desert parts of the public domain are converted to higher uses for the general benefit, was passed so that more Americans might have homes on the land. . . .

The obligations, and not the rights, of citizenship increase in proportion to the increase of a man's wealth or power. The time is coming when a man will be judged, not by what he has succeeded in getting for himself from the common store, but by how well he has done his duty as a citizen, and by what the ordinary citizen has gained in freedom of opportunity because of his service for the common good. The highest value we know is that of the individual citizen, and the highest justice is to give him fair play in the effort to realize the best there is in him.

The tasks this nation has to do are great tasks. They can only be done at all by our citizens acting together, and they can be done best of all by the direct and simple application of homely common sense. The application of common sense to common problems for the common good, under the guidance of the principles upon which this republic was based, and by virtue of which it exists, spells perpetuity for the nation, civil and industrial liberty for its citizens, and freedom of opportunity in the pursuit of happiness for the plain American, for whom this nation was founded, by whom it was preserved, and through whom alone it can be perpetuated. Upon this platform—larger than party differences, higher than class prejudice, broader than any question of profit and loss—there is room for every American who realizes that the common good stands first.

The Miracle of Irrigation

WILLIAM E. SMYTHE

THE ESSENCE OF the industrial life which springs from irrigation is its democracy. The first great law which irrigation lays down is this: There shall be no monopoly of land. This edict it enforces by the remorseless operation of its own economy. Canals must be built before water can be conducted upon the land. This entails expense, either of money or of labor. What is expensive cannot be had for naught. Where water is the foundation of prosperity it becomes a precious thing, to be neither cheaply acquired nor wantonly wasted. Like a city's provisions in a siege, it is a thing to be carefully husbanded, to be fairly distributed according to men's needs, to be wisely expended by those who receive it. For these reasons men cannot acquire as much irrigated land, even from the public domain, as they could acquire where irrigation was unnecessary. It is not only more difficult to acquire in large bodies, but yet more difficult to retain. A large farm under irrigation is a misfortune; a great farm, a calamity. Only the small farm pays. But this small farm blesses its proprietor with industrial independence and crowns him with social equality. That is democracy.

Industrial independence is, in simplest terms, the guarantee of subsistence from one's own labors. It is the ability to earn a living under conditions which admit of the smallest possible element of doubt with the least possible dependence upon others. Irrigation fully satisfies this definition.

The canal is an insurance policy against loss of crops by drought, while aridity is a substantial guarantee against injury by flood. Of all the advantages of irrigation, this is the most obvious. Scarcely less so, however, is its compelling power in the matter of production. Probably there is no spot of land in the United States where the average crop raised by dependence upon rainfall might not be doubled by intelligent irrigation. The rich soils

From *The Conquest of Arid America* (Seattle: University of Washington Press, 1969), 43–48. (Originally published 1899.)

of the arid region produce from four to ten times as largely with irrigation as the soil of the humid region without it. As the measure of value is not area, but productive capacity, twenty acres in the Far West should equal one hundred acres elsewhere. Such is the actual fact.

A little further on we shall see that not merely the quantity of crops, but their quality as well, responds to the influence of irrigation. We shall see how this art favors the production of the wide diversity of products required for a generous living. Certainty, abundance, variety—all this upon an area so small as to be within the control of a single family through its own labor—are the elements which compose industrial independence under irrigation. The conditions which prevail where irrigation is not necessary—large farms, hired labor, a strong tendency to the single crop— are here reversed. Intensive cultivation and diversified production are inseparably related to irrigation. These constitute a system of industry the fruit of which is a class of small landed proprietors resting upon a foundation of economic independence.

This is the miracle of irrigation on its industrial side.

As a factor in the social life of the civilization it creates, irrigation is no less influential and beneficent. Compared with the familiar conditions of country life which we have known in the East and central West, the change which irrigation brings amounts to a revolution. The bane of rural life is its loneliness. Even food, shelter, and provision for old age do not furnish protection against social discontent where the conditions deny the advantages which flow from human association. Better a servant in the town than a proprietor in the country!—such has been the verdict of recent generations who have grown up on the farm and left it to seek satisfaction for their social instincts in the life of the town. The starvation of the soul is almost as real as the starvation of the body.

Irrigation compels the adoption of the small-farm unit. This is the germ of new social possibilities, and we shall see to what extent they have already been realized as we proceed. During the first and second eras of colonization in this country the favorite size for a farm was about four hundred acres, of which from a fourth to a half was gradually cleared and the rest retained in woodland. The Mississippi Valley was settled mostly in quarter-sections, containing one hundred and sixty acres each. The productive capacity of land is so largely increased by irrigation, and the amount which one family can cultivate by its own labor consequently so much reduced, that the small-farm unit is a practical necessity in the arid region.

Where settlement has been carried out upon the most enlightened lines irrigated farms range from five to twenty acres upon the average, rarely exceeding forty acres at the maximum. It is perfectly obvious, of course, that a twenty-acre unit means that neighbors will be eight times as numerous as in a country settled up in quarter-sections—that where farms are ten acres in size neighbors will be multiplied by sixteen. Thus in its most elementary aspect the society of the arid region differs materially from that of a country of large farms. Eight or sixteen families upon a quarter-section are much better than no neighbors at all, but irrigation goes further than this in revolutionizing the social side of rural life.

A very-small-farm unit makes it possible for those who till the soil to live in the town. The farm village, or home centre, is a well-established feature of life in Arid America, and a feature which is destined to enjoy wide and rapid extension. Each four or five thousand acres of cultivated land will sustain a thrifty and beautiful hamlet, where all the people may live close together and enjoy most of the social and educational advantages within the reach of the best eastern town. Their children will have kindergartens as well as schools, and public libraries and reading-rooms as well as churches. The farm village, lighted by electricity, furnished with domestic water through pipes, served with free postal delivery, and supplied with its own daily newspapers at morning and evening, has already been realized in Arid America. The great cities of the western valleys will not be cities in the old sense, but a long series of beautiful villages, connected by lines of electric motors, which will move their products and people from place to place. In this scene of intensely cultivated land, rich with its bloom and fruitage, with its spires and roofs, and with its carpets of green and gold stretching away to the mountains, it will be difficult for the beholder to say where the town ends and the country begins.

This is the miracle of irrigation upon its social side.

Irrigation is the foundation of truly scientific agriculture. Tilling the soil by dependence upon rainfall is, by comparison, like a stage-coach to the railroad, like the tallow dip to the electric light. The perfect conditions for scientific agriculture would be presented by a place where it never rained, but where a system of irrigation furnished a never-failing water supply which could be adjusted to the varying needs of different plants. It is difficult for those who have been in the habit of thinking of irrigation as merely a substitute for rain to grasp the truth that precisely the contrary is the case. Rain is the poor dependence of those who cannot obtain the advantages of irrigation. The western farmer who has learned to irrigate thinks it would be quite as illogical for him to leave the watering of his potato-patch to the caprice of the clouds as for the housewife to defer her wash-day until she could catch rain-water in her tubs.

The supreme advantage of irrigation consists not more in the fact that it assures moisture regardless of the weather than in the fact that it makes it possible to apply that moisture just when and just where it is needed. For instance, on some cloudless day the strawberry-patch looks thirsty and cries for water through the unmistakable language of its leaves. In the Atlantic States it probably would not rain that day, such is the perversity of nature, but if it did it would rain alike on the just and unjust—on the strawberries, which would be benefited by it, and on the sugar-beets, which crave only the uninterrupted sunshine that they may pack their tiny cells with saccharine matter. In the arid region there is practically no rain during the growing season. Thus the scientific farmer sends the water from his canal through the little furrows which divide the lines of strawberry plants, but permits the water to go singing past his field of beets.

Plants and trees require moisture as well as sunshine and soil, and for three reasons: first, that the tiny roots may extract the chemical qualities from the soil; then, that there may be sap and juice; finally, that there may be moisture to evaporate or transpire from the leaves. But while all plant-life requires moisture, all kinds of it do not require the same amount, nor do they desire to receive it at the same time and in the same manner. Just as the skilful teacher studies the individualities of fifty different boys, endeavoring to discover how he may most wisely vary his methods to obtain the best results from each, so the scientific farmer studies his fifty different plants or trees and adjusts his artificial "rainfall" in the way which will produce the highest outcome. With the aid of colleges, experimental farms, and county institutes, wonderful progress has been made along these lines in recent years. This progress will continue until the agriculture and horticulture practised on the little farms of Arid America shall match the marvellous results won by research and inventive genius in every other field of human endeavor.

This is the miracle of irrigation upon its scientific side.

What Is Meant by Conservation?

LADIES' HOME JOURNAL

THE REQUEST FROM THE MOTHER:

I WONDER IF you ever realize, you who live and move in the big world of things, how little a woman like myself, living quietly up here, really knows of the great questions that seem so vital and throbbing to the country. "Where is your newspaper?" you will ask. But the newspaper is too verbose, to say nothing of the prejudiced writing and the previous knowledge its writers take for granted. I suppose I am like hundreds of women: I would keenly like to understand these great problems, but who is there to tell us, simply and clearly, and, don't forget, briefly?

What is meant by "Conservation"? Why do you say it should be more than a mere word to me? How does it affect me personally?

THE SON'S ANSWER:

I know, my dear Mother, just what's the matter. Conservation is a word so big and important-looking that it frightens you; but the idea behind it is as old as the Pharaohs, who in the seven plenteous years gathered corn to carry their people through the seven years of famine which followed. We have several words in common use with about the same meaning, like "economy," "thrift," "prudence," "forehandedness."

When you open a parcel from the store do you throw away the paper and string? Not a bit of it. You smooth out the paper and roll up the string, and lay both aside till you wish to wrap a package yourself. In the autumn you gather the seeds of your choicest flowers before burning up the dried stalks. The bones the butcher sends home with the meat you drop into the soup-kettle, and the surplus fat into the soap-can; the rain water from your

From *Ladies' Home Journal* 28 (November 1911), pp. 23, 95.

roof you catch for laundry purposes; your table refuse makes the pig and the chickens happy. So you have been practicing conservation all your life, doing on a small scale what the Government is beginning to do on a huge one, but you never spelled with a capital C. If the Government had begun as long ago as you did the people of the country would have been educated to the idea by degrees, just as you educated us boys not to be stingy, but to despise waste.

Now the Government is in a way the good mother of us all. She used to be rather easy-going, but she has lately come to realize that if she lets your generation and mine use up everything worth having there won't be enough for the next generation to live on. Where you save flower-seeds, therefore, she saves forests; where you store rain water for the washtub she fills reservoirs for irrigating desert lands and producing power for machinery.

Advantage Was Taken of the Government's Indifference

While the Nation was young and the supply of everything needful seemed inexhaustible people took advantage of the Government's indifference. Private parties fenced in and used for their own profit lands which belonged to all of us. If they wanted lumber for their houses, rails for their fences, or fuel for their stoves, they would cut down half a forest at a time; and whatever they could not use or sell they would leave to rot on the ground. They never bothered their heads to inquire where more wood was coming from when this was gone. As a result not only was the timber in some regions permanently exhausted, but the ground on which it had stood, being no longer shaded, parched under the strong sunshine and refused to bear any more trees or anything else.

That, Mother, is what went on for generations, and might have been going on still if ex-President Roosevelt had not interfered. He insisted that all good Americans must think as much of preserving their country for their children as of enjoying it themselves. . . .

Are you wondering why the name Conservation has been adopted for this movement, rather than Preservation? There is a nice shade of distinction between the two words. If you put a hindquarter of mutton into the icebox and keep it there for an indefinite period you *pre*serve it, certainly, but who gets any good from it? On the other hand, if you cut off from time to time what the family want to eat, chopping into hash the cooked meat left on the platter, turning the coarser fragments into a stew, and finally making soup of the bones, you *conserve* it: that is, you use every bit of it for the satisfaction of the family's hunger, but make it go twice as far as it would in the hands of a careless housekeeper. . . .

Our Forests and Mines Need Protection

About one-fourth of all the surface of the United States is timberland. We take out of this yearly some twenty-three billion cubic feet of wood. The yearly growth, however, is only about seven billion cubic feet; in other words, we are taking out more than three times as much as Nature is putting back. But that's not the whole story. Of the wood taken out not less than one-quarter is wasted, chiefly by carelessness in cutting or removing the logs. An average of fifty million dollars' worth is lost by fire every year, besides a good many human lives; and the young growth destroyed by fire is worth even more than the mature timber destroyed. Yet in spite of these fearful inroads on our supply we export a great deal of lumber, thus diminishing our own resources to make up the deficiencies of other countries.

No forest, once cut away, will grow again of itself as well as it did at first; so if we are going to replace what we remove we have got to plant new trees. To show you how little we are doing in this line all the forest land success-fully replanted in this country, from the beginning till the present day, could be set down inside of Rhode Island, the smallest State on the map, whereas we ought to have planted at least a hundred times that much. . . .

Try to Educate Your Neighbors

Now, Mother dear, don't lie awake nights worrying over what our descen-dants are going to suffer as the result of our neglect, or I shall be sorry I wrote you all this. There are more profitable occupations than worrying, and one is lending a hand promptly at stopping the leaks. Try to educate your neighbors a little by inducing them not to cut down a whole grove because they want to send a few logs to the sawmill or replenish the winter woodpile, but to select a tree here and there, where its removal will do least harm to the rest, and then to clear away the choppings and rubbish so that these will not furnish tinder for a forest fire. You can show the women of their families, also, how to make everything on the farm go farther by turn-ing it to more than one use. You can stir up the school-teachers in the neigh-borhood to impress upon the children the idea of taking care of a hundred things they are accustomed idly to destroy. Here is really the proper starting-point for the whole movement—with the children—for it is they who will have to pay the piper after we have had our dance.

If you wish to find out in what other ways a woman can help, send a postage stamp to the National Conservation Association at Washington, District of Columbia, and ask them to advise you. This reform is a big thing, and I only wish that every wide-awake woman like yourself could be enlisted in it.

The Other Side Of Conservation * oppose conservation

GEORGE L. KNAPP

FOR SOME YEARS past, the reading public has been treated to fervid and extended eulogies of a policy which the eulogists call the "conservation of our natural resources." In behalf of this so-called "conservation," the finest press bureau in the world has labored with a zeal quite unhampered by any considerations of fact or logic; and has shown its understanding of practical psychology by appealing, not to popular reason, but to popular fears. We are told by this press bureau that our natural resources are being wasted in the most wanton and criminal style; wasted, apparently, for the sheer joy of wasting. We are told that our forests are being cut at a rate which will soon leave us a land without trees; and Nineveh, and Tyre, and any other place far enough away are cited to prove that a land without trees is fore-doomed to be a land without civilization. We are told that our coal-mines would be exhausted within a century; that our iron ores are going to the blast-furnace at a rate which will send us back to the stone age within the lifetime of men who read the fearsome prophecy. In short, we are assured that every resource capable of exhaustion is being exhausted; and that the resource which cannot be exhausted is being monopolized. Owing to the singular pertinacity of the sun in lifting water to the mountain tops, and of the earth in pulling that water back to the sea, even the disciples of conser-

From *North American Review* 191 (1910), 465–81. Original footnotes have been removed.

vation by scareheads cannot say that in a few years we shall be a land without water-power. But they say the next worst thing. From official bureau and lecture platform, and from the hypnotized, not to say subsidized press, goes forth the cry that the water-power sites of the land are being hogged at a rate which will soon subject us all to the exactions of a cruel, soulless, grasping "power trust," the acme and consummation of all other trusts.

For all these evils which make the future a thing to dread, the remedy is "conservation." The "government," that potent "conjuh word" of civic atavists and political theologians, must stint its natural and proper tasks to engage in the regulation of this, that or the other industry, to "conserve" our resources. To "conserve" our timber, the wooded areas of the public domain, together with all lands touching on and appertaining to the wooded areas, and all other lands that might, could, would or should bear trees and don't, must be segregated from ordinary use and put under despotic control as "National Forests." To "conserve" our coal supply, the coal lands must be kept from passing into individual ownership, and operated, if at all, by persons who lease the privilege from the national government. To "conserve" our water-power, the power sites must be treated as the coal lands, and developed, if at all, as leaseholds. In a word, the Federal Government must constitute itself a gigantic feudal landlord, ruling over unwilling tenants by the agency of irresponsible bureaus; traversing every local right, meddling with every private enterprise, which seems to stand in the way of the sacred fetish of "conservation."

Only by such drastic means, we are told, can the rights of the people be protected, and the continued prosperity of the nation be assured. So persistently and adroitly has this view been urged by this press bureau, that millions of people wonder, in their innocence, why any one should object to so needful and righteous a work. Acting doubtless on the suggestion of the founder of the Ananias Club, the conservation press bureau has impugned the motives of all who disagree with it. If one objects to the inclusion of non-forest land within forest reserves, he is ranked forthwith as a would-be robber of the public domain. If he doubts the propriety of the Federal Government setting up in business as a professional savior from imaginary ills, he is an "individualist"—that being the bitterest term of reproach in the "conservation" vocabulary. If one objects to the leasing of the coal lands, he is plainly an undesirable citizen of some sort; and if he declares the proposed "conservation charge" for water-power to be both unconstitutional and silly, he is marked at once as an emissary of that fearful "power trust" which is so unconscionably long a-borning.

Notwithstanding the ban thus threatened, I am going to enter the lists. I propose to speak for those exiles in sin who hold that a large part of the present "conservation" movement is unadulterated humbug. That the modern Jeremiahs are as sincere as was the older one, I do not question. But I count their prophecies to be baseless vaporings, and their vaunted remedy worse than the fancied disease. I am one who can see no warrant of law, of justice, nor of necessity for that wholesale reversal of our traditional policy which the advocates of "conservation" demand. I am one who does not shiver for the future at the sight of a load of coal, nor view a steel-mill as the arch-robber of posterity. I am one who does not believe in a power trust, past, present or to come; and who, if he were a capitalist seeking to form such a trust, would ask nothing better than just the present conservation scheme to help him. I believe that a government bureau is the worst imaginable landlord; and that its essential nature is not changed by giving it a high-sounding name, and decking it with home-made haloes. I hold that the present forest policy ceases to be a nuisance only when it becomes a curse. Since that forest policy, by the modest confession of its author, is set forth as the model to which all true "conservation" should conform, I shall devote most of my attention in this paper to the much-advertised "National Forests" and their management. . . .

The terrors from which "conservation" is to save us are phantoms. The evils which "conservation" brings us are very real. Mining discouraged, homesteading brought to a practical standstill, power development fined as criminal, and, worst of all, a Federal bureaucracy arrogantly meddling with every public question in a dozen great States—these are some of the things which result from the efforts of a few well-meaning zealots to install themselves as official prophets and saviors of the future, and from that exalted station to regulate the course of evolution.

It is no more a part of the Federal Government's business to enter upon the commercial production of lumber than to enter upon the commercial production of wheat, or breakfast bacon, or hand-saws. The judiciary committee of the Sixtieth Congress, reporting on the proposed Appalachian reserve, declared that the sole ground on which Congress could embark in the forest business was the protection of navigable streams. Will any one pretend that a forest reserve on the crest of the Rocky Mountains, with the nearest navigable water a thousand miles away, can be brought under this clause? Even on the Pacific slope, I have not heard that the lumber mills of Washington have seriously impaired the navigability of Puget Sound; nor that the Golden Gate would shoal up if the cutting of timber in the Sierras were unchecked. And will the champions of "conservation" claim that the Federal Government has greater rights and powers in the newer States than in the older ones?

But the public lands belong to the whole people. Undoubtedly; but in what sense do they so belong? As a landed estate, from which to draw rentals, or as an opportunity to be used? Which interpretation of this ownership has prevailed in the past? Which doctrine caused the settlement of a region as large as half Europe within the lifetime of a single generation? And passing this larger aspect of the question, if the "people" do own the public lands, and especially the "National Forests," in the sense of being possessors of a rentable estate, are they quite sure that it will pay to treat that estate in that fashion? The total receipts from the "National Forests" in 1908 were $1,842,281.87. The expenditures for the same year were $2,526,098.02, leaving a deficit of $683,816.15. If the "people" really want that deficit and would feel robbed without it there might be less bothersome ways of supplying their need than the maintenance of a Federal bureau. It might be cheaper to sell the estate on reasonable terms and trust to the patriotic endeavors of Congress to provide the indispensable deficit.

Our natural resources have been used, not wasted. Waste in one sense there has been, to be sure; in that a given resource has not always been put to its best use as we now see that use. But from Eden down, knowledge has been the costliest thing that man could covet; and the knowledge of how to make the earth best serve him seems well-nigh the most expensive of all. But I think we have made a fair start at the lesson; and considering how well we have already done for ourselves, the intrusion of a Government schoolmaster at this stage seems scarcely needed. The pine woods of Michigan have vanished to make the homes of Kansas; the coal and iron which we have failed—thank Heaven!—to "conserve" have carried meat and wheat to the hungry hives of men and gladdened life with an abundance which no previous age could know. We have turned forests into villages, mines into ships and sky-scrapers, scenery into work. Our success in doing the things already accomplished has been exactly proportioned to our freedom from governmental "guidance," and I know no reason to believe that a different formula will hold good in the tasks that lie before. If we can stop the governmental encouragement of destruction, conservation will take care of itself.

To me the future has many problems but no terrors. I belong to the generation which has seen the birth of the electric transformer, the internal-combustion engine, the navigation of the air and the commercial use of aluminum, and I quite decline to worry about what may happen "when the world busts through." There is just one heritage which I am anxious to transmit to my children and to their children's children—the heritage of personal liberty, of free individual action, of "leave to live by no man's leave underneath the law." And I know of no way to secure that heritage save to sharply challenge and relentlessly fight every bureaucratic invasion of local and individual rights, no matter how friendly the mottoes on the invading banners.

Another National Blunder * oppose conservation

H.J.M. MATTES

WHY DO THE common people object to reserves? It is a question between kinds of government—popular government and monarchical government. Under the former the people are supreme; under the latter they have no rights until they are granted them by the supreme ruler. Outside reserves, the common people legally help themselves to the timber; inside, they must first ask permission from some representative of the supreme ruler, the Honorable Secretary of the Interior. Talk about imperialism!

But should the people be allowed so much liberty? It is wise that they should be. Timber is one of the vital necessities of life—for fuel, for buildings. The East has allowed all its coal to come under private ownership, and with a country full of coal they are having a famine. Would it be wise to put all our timber in charge of one man? Even the common law protects the people's right to free timber. In Colorado last year a gardener dug a tree from one man's yard and planted it in another's as a gift. In the absence of

From *Forestry and Irrigation* 9 (May 1903), 258–62.

any statute, common law was applied, and it was found that there was no theft, no trespass, and not even malicious mischief.

What then becomes of the great cry of "stealing government timber"? In a monarchy it might be stealing—in some very tyrannical monarchy—and has been punished by death, but in America it has always been our right even as we have a right to the land, to the rain that falls, and to the air we breathe. There is an old national statute making it a trespass, but other laws have been passed restoring the right to the people while keeping it from the corporations.

"How, then, shall we protect our timber?" It does not need protecting. After building up the mightiest nation on earth, one-third of our land still grows timber; and while that may not be enough for the future, we can draw on the governor of Canada, who has discovered the largest forest of the world in his dominions, 4,000 miles long, 700 miles wide, and offers to supply us all we can use for a century if we will merely take off the tariff.

"But the sawmills are slaughtering it." The poor sawmills! They have borne more abuse than the early Christians. They only cut the big, ripe trees. It was the farmers that girdled the trees and made bonfires of them, and then pulled everything up by the roots. It was the farmers that turned the impervious subsoil on top of the spongy mold and caused the freshets and dried up the springs. Yet [President Grover] Cleveland devoted his first term trying to annihilate them. With more than Christian meekness they said nothing and went on sawing wood. The sawmill men appreciate their high vocation. When they stop sawing the nation will stop growing, and civilization will come to a halt and the world start back toward chaos again. They even get blamed for all the bad weather, but if the chief of our Weather Bureau knows anything about climate, timber has no appreciable effect upon it. . . .

Let the sawmills alone. Drop the timber question. Those two great world builders, the settler and the sawmill man, have marched across the American continent hand in hand and built up the mightiest nation on the earth in one of the greatest wildernesses; and they have much work ahead of them. Let them alone.

What, then, shall we do with the Forestry Bureau? If these "scientists" will not keep out of mischief and let the West alone, abolish it. Look at their ignorant interference. They are telling us that the little sheep pack the ground so that the rain will not sink into our gravelly soil, while the big cows and horses do no harm. They are telling the sawmill men to cut this tree and not cut that one, while we have to make the kind of lumber the settlers need and choose the trees that will make it. We know our business. They are telling us to burn the offal, when a lighted cigar will easily start a fire in our tinderbox forests. And the Interior Department tries to compel the frequenters of the reserves to do these things or go to prison. . . .

Shall this nation go on growing, or shall we go on making forest reserves and forever stop its growth? Shall the people have their natural rights restored to them and preserved, or shall the government cater to the spirit that depopulated Europe and built up America? Shall our pioneers be our natural heroes, or the forestry faddists and "scientists," the nation's pets? Read "Forest Law," in Encyclopedia Britannica, and compare those laws with the reserve laws shown in the Land Office circulars, and you will agree with me that forest reserves and human liberty cannot stand upon the same ground. One or the other must go down.

Part 2

PERSPECTIVES ON WILDLIFE

CONSERVATION

Although Roosevelt, Pinchot, and others worked hard to develop a political consensus for their conservation policies, a wide range of opinions persisted. This section provides a sampling of the diverse perspectives on one important conservation issue—the protection of wildlife.

George Bird Grinnell had a long and active career in wildlife protection, most importantly serving as editor of *Forest and Stream,* which he began publishing in 1873. In this excerpt he gives a retrospective view of the movement, with a particular emphasis on the role of hunters in conservation. Grinnell supported the conservation of wildlife for its continued availability for sportsmen, like himself, and toward this goal he helped found the Boone and Crockett Club in 1888.

The second piece comes from Mabel Osgood Wright, a cofounder and first president of Connecticut's Audubon Society, and an editor of the Audubon Societies' *Bird-Lore.* While Grinnell's writing emphasizes the importance of male leaders in the movement, Wright describes the importance of community activism, particularly in maintaining long-term vigilance. With so many of the prominent men focused on federal legislation, Wright reminds us of the importance of state and local activities to the broader movement.

The third piece comes from William T. Hornaday, a well-known naturalist, author, and the director of the New York Zoological Society. Among Hornaday's many publications was *Our Vanishing Wild Life,* a book-length treatment that blames both market and sport hunters for the decline in wildlife. Last, in a memoir written around 1935 and published in 1958, David Merrill reflects on his early life in the Adirondacks. Merrill describes how his family made a living off the land in the late 1880s; they did so in a region changing due to an influx of tourists and tightening natural resource regulations. Merrill reveals his family's negative impression of conservation.

American Game Protection: A Sketch

GEORGE BIRD GRINNELL

GAME PROTECTION IN North America has passed through three stages—has been influenced and guided by three successive motives. The first of these was selfish—in which sportsmen wished to lessen the killing of game in order that sufficient might be left alive to furnish abundant sport for themselves. This motive governed for nearly a generation. The second motive was sentimental, where a large and ever increasing number of people were interested in wild life protection because these living objects are beautiful to look at and ought to be preserved so that we and our successors may have the pleasure of seeing them. The third motive for protection is economic, and considers these wild things as assets which possess a tangible value to the community and so are worth preserving; with the further thought that they have been given to us as trustees to hold for those who are to come after us. This view holds that money expended in preserving them from destruction is in the nature of an insurance premium. Of these three motives, the third—the economic view—is constantly gaining strength.

Experience has shown that if these wild things are utterly destroyed, a later generation will feel that it must replace them. This replacement is often impossible, but even if it can be accomplished, the task is one of time, difficulty and great expense. It is far more economical to spend today a little money to keep these living things in existence than to replace them at a later time—or to try to replace them and fail to do so.

That the Congress has seen light in this matter is shown by its action in establishing national parks, forest reservations, game preserves and national monuments. . . .

From *Hunting and Conservation: The Book of the Boone and Crockett Club* (New Haven, Conn.: Yale University Press, 1925), 201–57.

Efforts for Preservation

In early colonial times the new country swarmed with wild creatures; the lands and waters were crowded with birds and mammals to the limit of their support. Game was free to all, liberty to take it at will was regarded as a universal right. For a long time there seemed to be no necessity to restrict this right of capture. When, later, attempts were made to limit the privilege of taking game, these were resented as efforts to deprive the people of something that belonged to them. Legislation to protect some forms of game was initiated in the Colonies in the seventeenth century, but the laws then placed on the statute books were nothing more than an inheritance from English ancestors and were never enforced. It was only in the nineteenth century that sportsmen and naturalists began to awaken to the danger of extermination which threatened many of our wild species.

The first active steps looking toward game protection in the United States were taken in New York early in the year 1844, when was established the New York Association for the Protection of Game, which thus has now behind it a career of three-quarters of a century. For many years it labored alone, supported by its own enthusiasm and its own contributions. Its membership included some of the best New York sportsmen and it gradually built up, first among residents of New York City and later in other communities, a spirit for game protection which has increased and broadened, with results which we see today. . . .

The Association naturally devoted itself to work in New York State, and after a time secured from the Legislature the passage of its draft of a law by which the possession of game out of season was made not less a violation of the law than the illegal taking of game. It was this enactment that made game law enforcement possible in this country. . . .

The proportion of sportsmen in the community was small and little interest was felt by others; and since in those days, as now, the absorbing question to most people was how to earn a living or to make more money, the men who enjoyed shooting or fishing were looked upon as idlers and ne'er-do-wells, for it was thought that these pursuits were mere excuses for laziness—the avoidance of work. It was not until about the year 1880 that the status of the shooter and the fisherman began to rise in public estimation, his motives to be comprehended, and the old prejudice against gun and rod and dog to be given up. . . .

About that time journals were established which were devoted to shooting and fishing and natural history, and to the increase of the supply of fish and game. At first, undertaking a novel task, they worked inefficiently. Nevertheless, little by little they extended their sphere of influence, reached sportsmen in many widening fields, and, preaching earnestly the doctrine of protection, sought to make new converts and to give encouragement and assistance to those already interested in the movement. As the journals of sport discussed it more and more, the daily newspapers began to notice the subject, and at length wrote of it from their own point of view. In a few years a considerable propaganda was set on foot. The protective movement was taken up by the anglers, and much attention was given to fish culture, which has since developed into so important a factor in the national food supply. Closely akin to the preservation of game was the preservation of forests, in which for a number of years a few men had been working against constant discouragement. . . .

Especially destructive of wild life was the practice of shooting wild fowl and shore birds at all seasons, the work of the market hunter, who shot as long as he could sell his game, and that of the economic importance of the principle for which they were striving and having put their hands to the plow did not look back. Gradually the seed began to promise a crop. As the years passed, more and more people became interested and the ranks grew stronger, until at last the body of wild life protectors became a force in the land and today constitute an element which is sure of a hearing whenever it is demanded. The public has been educated to a point where it understands that conservation of wild life is for the general good. . . .

Temperate America has not been very long settled as world history goes. A little more than three hundred years covers the period. Yet in that time a dozen species of birds and mammals have become extinct—some of them in our own time—and others are on the way to extinction. It should be the concern of each one of us to put off that day as long as we can.

Migratory Bird Protection

The final achievement in the long struggle for wild bird protection was accomplished in 1918. This was the enactment of legislation by which migratory birds were placed under control of the Federal Government. The United States now controls the killing of birds which spend one season of the year in the North and another in the South, and in their journeyings may pass over half a dozen states or provinces. Not long after the enactment of the Federal Law, a treaty was concluded between the United States and Great Britain [which controlled Canada] for the protection of migratory birds in the two jurisdictions.

The law and the treaty provide for the federal protection of all migratory non-injurious birds. So long as law and treaty shall stand and are enforced those species will be safe. . . .

Federal Bird Refuges

In 1903, Theodore Roosevelt set aside the first federal bird reservation on Pelican Island in Florida, and during the following years he established almost fifty more, setting an example which, in a less degree, has been followed by later presidents, so that there are now over one hundred bird reservations, in about twenty states and two territories. . . .

All this has come about in little more than a generation. As already suggested, its beginning was set on foot many years ago by people who, while they had a general idea of the results they hoped for, did not then know how these results were to be attained, and had no slightest conception of the final outcome of their efforts. Their confidence that they were right and their faith in the good sense of the American people were their only support. They worked along from day to day, blindly and gropingly; but all the time they were learning, and the call they sounded appealed to many people. Gradually the work was taken up by thousands upon thousands of earnest men and women, many of whom came to have a splendid enthusiasm for a cause whose meaning little by little became more and more clear. Converts were made and young people grew up who became devoted to some phase of protection. Of these, not a few possessed energy and brains, and forged to the front and often bore the burden of the battle.

Since in most states there exists a strong feeling in behalf of wild life protection, since the chiefs of three important government bureaus appreciate the importance of recreation and of the protection of natural things, and since also scattered throughout the country are many associations and individuals devoted to such protection and actively watching Congress so as to forestall the frequent commercial attacks on our national parks, our national forests and our wild life, we see already the dawning of a day when, in suitable situations, the forests, the birds and the animals so ruthlessly swept away in the past may be in a measure reestablished and, within proper limits, may be preserved for the benefit of future generations.

Keep on Pedaling!

MABEL OSGOOD WRIGHT

TEN YEARS AGO, when the world and his wife were striving to master the vacillating bicycle, the constant cry of the perspiring instructor who ran beside was, "Keep on pedaling; if you stop you're a goner!"

This concise if inelegant advice applies to many things besides wheeling—and especially to the work of bird protection. At the present moment thirty-odd Audubon Societies are more or less securely mounted and started upon the right road; but if, in addition to "keeping on pedaling," they do not look both right and left as well as ahead, an upset will speedily follow.

We are all prone to overestimate the importance of initial effort, whether it be in mastering a horse, a wheel, or in organizing a new movement. Of course, in order to have a cooperative society there must be organization, but the organization should be regarded only as a platform upon which the members may stand united to work intelligently for reaching an end, not as the end itself.

When you often hear some one say, "Oh, yes, birds are being protected in our state, there is nothing to worry about there; we have just started an Audubon Society," as if a declaration was all that was necessary, you will understand the necessity of the injunction to "keep on pedaling."

As the societies have, for their motto, The Protection of Birds, so, if they would work with any hope of success, they must stand upon one platform, Public Education, and public education is something that is as endless as the race itself. It is true that public education in a general sense has obtained long enough in this country to be regarded as an inalienable right; but until the lesson of protection of all forms of harmless and useful animal life is so well learned as to become part of the heredity of coming generations, any relaxing in vigilance in the different branches of protection will be fatal to

From *Bird-Lore* 6 (1904), 33–34.

the whole cause; and for this reason every society should have special committees ever on the watch for pitfalls. . . .

The committee on Bird Study in Schools should be composed of people of both sexes who not only have a knowledge of the child-mind, but of the amount of work already obligatory in the different grades; then less fault will be found with teachers for "not showing interest" and greater results will follow.

The Legislative Watch-Out Committee should be composed of the shrewdest men available, with a knowledge of state politics. If one is a lawyer all the better; he may save the rest from running their heads into legal nooses at times when they most need them. A good committee of this sort will often engage the interest of many men who would otherwise see no work for themselves in an Audubon Society, not appreciating the value of a promise "not to wear the feathers of song-birds for decorative purposes."

On the other hand, a large mixed body—drawn from widely different corners, the more so the better, can be organized under the head of Committee for the Posting of the Laws. This vast work cannot be done by a few, and the work rivals in importance the making of the laws themselves; but if fifty or one hundred persons in each state could be relied on to undertake the matter, these in turn may employ local help until the chain is complete. How much more interesting would be the oftentimes perfunctory annual meeting if these three before-mentioned committees brought in full reports!

One of the beauties of a progressive country is that where everything moves nothing can be fixed; it must either go forward, backward, or drop out. Part of legal prerogative is that any legislative session may untie the knots made apparently firm by another, so the Watch-Out Committee must be never-ending. . . .

As we hope that bird-life may never be extinct, on our continent, at least, so must the work of the Audubon Societies be perpetual.

All cheer for 1904, good friends; hold your handle-bars firmly, mind sharp—legislative— curves, and above all, "keep on pedaling."

Our Vanishing Wild Life:
Its Extermination and Preservation

WILLIAM T. HORNADAY

✻ Preservation

✻ ZOO

THE WRITING OF this book has taught me many things. Beyond question, we are exterminating our finest species of mammals, birds and fishes *according to law!*

I am appalled by the mass of evidence proving that throughout the entire United States and Canada, in every state and province, the existing legal system for the preservation of wild life is fatally defective. There is not a single state in our country from which the killable game is not being rapidly and persistently shot to death, legally or illegally, very much more rapidly than it is breeding, with the extermination for the most of it close in sight. This statement is not open to argument; for millions of men know that it is literally true. We are living in a fool's paradise.

The rage for wild-life slaughter is far more prevalent to-day throughout the world than it was in 1872, when the buffalo butchers paved the prairies of Texas and Colorado with festering carcasses. From one end of our continent to the other, there is a restless, resistless desire to "kill, *kill!*"

I have been shocked by the accumulation of evidence showing that all over our country and Canada fully nine-tenths of our protective laws have practically been dictated by the killers of the game, and that in all save a very few instances the hunters have been exceedingly careful to provide "open seasons" for slaughter, as long as any game remains to kill!

✻ *And yet, the game of North America does not belong wholly and exclusively to the men who kill! The other ninety-seven per cent of the People have vested rights in it, far exceeding those of the three per cent. Posterity has claims upon it that no honest man can ignore.* ✻

I am now going to ask both the true sportsman and the people who do not kill wild things to awake, and do their plain duty in protecting and preserving the game and other wild life which belongs partly to us, but

From *Our Vanishing Wild Life* (New York: Charles Scribner's Sons, 1913), ix–x.

chiefly to those who come after us. Can they be aroused, before it is too late?

The time to discuss tiresome academic theories regarding "bag limits" and different "open seasons" as being sufficient to preserve the game has gone by! We have reached the point where the alternatives are *long closed seasons or a gameless continent;* and we must choose one or the other, speed- ily. A continent without wild life is like a forest with no leaves on the trees.

The great increase in slaughter of song birds for food, by the negroes and poor whites of the South, has become an unbearable scourge to our migra- tory birds,—the very birds on which farmers north and south depend for protection from the insect hordes,—the very birds that are most near and dear to the people of the North. *Song-bird slaughter is growing and spreading,* with the decrease of the game birds! It is a matter that requires instant attention and stern repression. At the present moment it seems that the only remedy lies in federal protection for all migratory birds,—because so many states will not do their duty.

We are weary of witnessing the greed, selfishness and cruelty of "civi- lized" man toward the wild creatures of the earth. We are sick of tales of slaughter and pictures of carnage. It is time for a sweeping Reformation; and that is precisely what we now demand.

I have been a sportsman myself; but times have changed, and we must change also. When game was plentiful, I believed that it was right for men and boys to kill a limited amount of it for sport and for the table. But the old basis has been swept away by an Army of Destruction that now is almost beyond all control. We must awake, and arouse to the new situation, face it like men, and adjust our minds to the new conditions. The three million gunners of to-day must no longer expect or demand the same generous hunt- ing privileges that were right for hunters fifty years ago, when game was fifty times as plentiful as it is now and there was only one killer for every fifty now in the field. . . .

The Education of a Young Pioneer
in the Northern Adirondacks

DAVID SHEPARD MERRILL

IN THE NORTHERN Adirondacks, in the counties of Franklin and Clinton, lie the beautiful Chateaugay Lakes, comprising a navigable water course of approximately twelve miles in length, consisting of the Lower Lake, the Narrows and the Upper Lake. Near the north end of the Lower Lake on the east side was a small cottage where my memory begins. At this time my father had bought land and built a log cabin at the north end of the Upper Lake. This cabin was about seven miles distant from our home and was reached by rowboat up through the Lower Lake and the Narrows. My father had four rowboats and his work in summer was mostly in taking sportsmen and pleasure seekers up to his cabin at the Upper Lake where there was good fishing and hunting.

During a few weeks in Spring and Fall he caught fish with a seine and sold them in Malone and Chateaugay and to farmers along the road. There were four children in our family, Minnie, Watson Paul, David Shepard (myself) and Charles Ernest. When I had arrived at the age of seven and brother Watson was nine we were allowed to go with father to the Upper Lake to help haul the seine. This was a great treat to us and good help for father, and later when I was ten and brother Wat was twelve we were able to do this work without father's help.

Seine fishing was legal at this time providing we caught only whitefish. It was very seldom that we caught any trout. Seine fishing had to be done in the night as the whitefish did not run on the sand flats in day time. Our seining outfit consisted of eighty rods of three eights rope and a seine ten rods long, four feet wide at the ends and twelve feet wide in the middle, strung on quarter inch ropes, one with leads the other with floats, with hard wood staff four feet long at each end. To "ship" the outfit, we put one forty rod section of the rope into the boat near the stern, then the seine on top

From *New York History* 39 (July 1958), 238–55.

of it, then the other forty rod section of rope on top of the seine. To "put out" the seine, we attached the end of the top rope to the shore, then rowed out at right angles to shore till the top rope was run out, then turned and rowed parallel with the shore till the seine was run out, then turned toward shore, landing at a point twelve rods from the point where the other rope was attached to the shore. Then we began hauling on both ropes and as the seine approached shore, we came in till the two haulers were ten feet apart. Then the middle or wide part of the seine was hauled out onto the beach.

The mesh of the seine was inch and a quarter so we didn't get any small fish. With favorable weather for a good run we would get as many as fifty at one haul. It takes an hour to make a haul and we had to change ground for each haul. The catch averaged about a pound each. . . .

In the good old days we went deer hunting, primarily, to get food for the household, and like the Indians we did not pay much attention to the game laws of which there were not very many at that time. However there was a universal code that deer should not be disturbed while "yarding," or in the breeding season, and this applied to game birds as well. Of course we kids learned to handle guns at an early age. I was eight and Wat was ten when we got our first deer. . . .

I believe it was in the fall of 1875, one bright morning in September a party came from Malone for a one day hunt. Father took the dogs in on the west side and told me to watch at Pople Point. By eleven o'clock I had not heard a sound of the dogs, and started to pull over home. When half a mile from shore I saw a huge buck swimming towards me from the east. I waited till he was within fifty yards or so, when he started in the opposite direction. I started after him and had a hard pull to get near enough to shoot. I had father's double barrel combination, shot and rifle, muzzle loader. I pulled on him with the rifle and got him the first shot. After bleeding him I tried to get him into the boat but he was too heavy for me, so I started to tow him home when Nat Collins came along and helped me get him into the boat. I took him home, had a bite to eat, reloaded the gun and went back, stopping at the beach in Shattuck's Bay to get a drink from the little brook there. As I started to shove off, a big doe came across the beach and started swimming straight out into the lake. I soon caught up to her and gave her the charge from the shot barrel. Two shots with the old muzzle loader and two deer, a very good day's work for a boy of fourteen. However, where I had one lucky day hunting there would be a dozen days without anything to show for it. Many times we would hunt a full week and not get a single deer. I learned a lot about the habits of deer and other wild animals. For some unknown reason they will migrate from one section of the woods to another, probably in search of better feeding conditions. . . .

We had such a diversity of employment that there was no need for vacations. Each change was a vacation, and there was hardly a dull moment in the whole year. However, at the closing weeks of winter we became anxious to get into the sugar bush, and at the close of the sugar season we were glad to get out and start spring seine fishing and the farm work. Only for the fact that our education was sadly neglected, it would have been an ideal life. Our work out in the open gave us healthy bodies and strong muscles, yet later in life we found ourselves handicapped by the lack of schooling. Later, when new game laws were made, prohibiting seining and netting in waters inhabited by trout, we had to quit this form of fishing. It was a great disappointment to us and we fought it for a while and we boys were in court on two or three occasions as witnesses against father. In court we boys managed to avoid answering the lawyers' questions by stating that the answer might incriminate ourselves. Father believed that netting with ⚡ nets of proper size mesh was the right way to catch fish as this would take only the large ones, the smaller ones going through the mesh. On the theory that the fish were put here by our Creator for food, and not for sport, the theory is correct. The sportsman's catch of trout is mostly small ones which would go through the mesh of a net and live to grow to a decent size.

My father did not believe it was right to catch fish for sport, but for food only. And the right method for fishing was with a net that would take only the larger fish. The hook and line method for fishing takes mostly small fish, and in this way you are killing some six fish to get a pound of food, whereas with the net you kill one fish and get two pounds of food. However if the fish were put here for the purpose of providing sport for pleasure seekers, the net would not be the proper equipment for they prefer to torture the fish for the fun they get out of it. A fish as well as a man has a sense of feeling and suffers after being hooked, and the longer this period of torture can be extended, the more fun the so-called "sportsman" can get out of it. I believe God created the fish for food for mankind and not to satisfy a depraved and perverted mind. In waters where netting is legal there is always plenty of fish; on the other hand, in waters where netting is prohibited, there is always a scarcity, even when restocking is done continuously. This is positive proof that netting increases the supply of fish. Does this not also prove that hook and line fishing is destructive? . . .

Part 3

THE UTILITY OF "CONSERVATION"

A s the conservation movement gained broader support, the utility of the word "conservation" spread as well. By the 1910s, reformers interested in solving any number of problems adopted the language of conservation and related the themes of that movement to their own interests. Here, famed labor leader Samuel Gompers, president of the American Federation of Labor, offers his support to conservation. In an editorial for the union magazine he edited, *American Federationist,* Gompers connects conservation of natural resources with the conservation of human resources. In a critique of American capitalism, Gompers associates the wastefulness of unemployment caused by an economic downturn with the wastefulness of "marauders" of natural resources. Reflecting the optimism and democratic fervor of William Smythe, Gompers predicts that energetic government will create both greater and better distribution of wealth.

J. Horace McFarland, the most prominent member of the City Beautiful Movement, argued that conservation for productivity would not be sufficient. According to McFarland, success could only come when conservation considered beauty, healthfulness, and habitability too, all of which were mentioned by the National Conservation Commission. In this essay, excerpted from the progressive magazine, *The Outlook,* McFarland addresses beauty and conservation, focusing on American cities, and particularly on the ugly waterways that ran through them. Why couldn't conservation protect visual resources as well as productive resources?

Following similar themes, the prominent reformer Mary Ritter Beard touted the great accomplishments of women working to improve their cities, particularly their attempts to rescue terribly disorderly, immoral, and ugly places within America's municipalities. In her chapter entitled "Civic Improvement," from a lengthy book, *Woman's Work in Municipalities*

(1915), Beard traces the work of women in a number of areas related to city planning, simultaneously articulating many of the themes that the City Beautiful Movement shared with the broader conservation movement.

In 1909, Yale economist Irving Fisher connected conservation and human health. Writing for the National Conservation Commission report, Fisher argues that conserving health would be essential for improving national efficiency. Here he references recent scholarship concerning human efficiency and health. Similarly, in *Conservation By Sanitation* (1911), well-known home-economics pioneer Ellen H. Richards provided a book-length treatment of air and water sanitary practices, and the excerpts below emphasize the links Richards makes between conservation and sanitary improvements in cities. They also reveal how regularly those interested in environmental improvements could speak of both conservation and preservation in the same breath.

Conservation of Our Natural Resources

SAMUEL GOMPERS

IT IS A MATTER of profound interest and gratification to the American people that the convention of governors of states, the forestry, irrigation, and waterway engineering experts and others who have given the public weal their study was called together by the Chief Executive of the nation.

These eminent citizens are gathered in obedience to a call, the inspiration of which strike the key-note of the nation's future policy in the field of civic betterment. It is the extension of the new school of political economy. It is in the nature of the great stewardship that underlies the brotherhood of man. No more noble incentive to that end can be imagined than is to be found in the impulse that prompts wise and far seeing statesmanship to build and preserve for the future. Happily, too, this convention will act as a check on the marauding instinct so flagrantly exercised in the exploitation of the nation's natural resources by men whose actions have hitherto been sanctioned by law. In respect of waste and extravagance in the economic sense, these marauders have placed the American Republic in a situation unparalleled in economic conservation among the nations. In one item alone, that of fuel, it is figured out by one of the experts attendant upon this convention that 200,000,000 tons of coal are wasted every year in the mining processes of the nation, which is equal to $200,000,000, every ton of coal being worth a dollar at the mines. Add to this the colossal waste in the exploitation of timber lands, water power, and the like, and we have some faint conception of the load our economic energies are carrying. . . .

Perhaps the greatest form of waste from which we suffer at this time is the waste involved in the unemployment of immense numbers of our people, and this waste is due to no fault of the working people. Without affirming to whom is traceable the blame of the present condition of unemployment, no one can truthfully charge that the cause can be laid at the door of the

From *American Federationist* (July 1908), 532–36.

working people of our country. It is perhaps the severest commentary upon the intelligence and understanding of economic and social conditions that there are at this time about two millions of our working people vainly seeking the opportunity of working and earning their bread by the sweat of their brows. Apart from the demoralizing influence of such a state of affairs, a mere statement of the material loss may not be uninteresting.

It is now seven months since last October when the panic was thrust upon our people. Counting 25 normal working days for each month, we have a total of 175 working days, and giving the conservative sum of $2.50 wealth produced daily for each worker, there has been a waste or loss of $875,000,000 of wealth which could and would have been produced but for uneconomic methods.

In our uneconomic methods no accounting is here given of the myriads of workers whose bodies are maimed or whose lives are destroyed in industry and commerce by ignorance, incompetence or greed. What is more wasteful, what is more the antithesis to the conservation of our natural resources? . . .

When we see the great cities of our country and their environments stretching away into their then suburban population, and all the intervening territory a bounteous and marvelous reservoir that shall not only house, feed, clothe, and educate the people of our mighty republic itself, but which shall become a gigantic mill to grind our products and commodities for which the whole world will bid, we shall attract to our shores the vine and the flower of the world's civilization.

We shall in truth become the asylum of the oppressed, and as we now stand for political and religious liberty, so shall we become the exemplar and defender of industrial freedom.

When we have all of these advantages, the labor energy of the republic will dominate the world.

For it must be borne in mind that labor energy is the foundation of all wealth. Nothing is accomplished without labor. . . .

When we are expanded and developed, our natural resources wrested from the hands of heedless marauders and conserved as the nation's patrimony; when our great natural waterways are connected with canals; when our denuded watersheds are rehabilitated and made verdant and fruitful; and when the nation through the people speaks, the working men and the working women, who in reality do everything that is done, will take their place in freedom. . . .

Shall We Have Ugly Conservation?

J. HORACE MCFARLAND

"LET US CONSERVE the foundations of our prosperity," concludes the noble Declaration of the Governors assembled at the memorable White House Conference of May, 1908. The same Declaration—a new Declaration of Independence, as vitally related to the future of the United States as was the great document of 1776—among its definite recommendations insists "that the beauty, healthfulness, and habitability of our country should be preserved and increased." This insistence stands equivalent to the accompanying urgency that we must stop soil erosion, reclaim waste lands, control waters, perpetuate the forests, and use sensibly the minerals of our country.

Four great sections of a National Conservation Commission were appointed, to deal respectively with waters, forests, soil, and minerals, in harmony with the suggestions of the President and the Governors. These Commissions have met, labored with public-spirited zeal, and their reports include an inventory of our natural resources—with one exception.

The duty of considering "the beauty, healthfulness, and habitability of our country" was not assigned to any of these sections, though vitally connected with the work of each of them. Apparently, the recommendation of the Declaration concerning health and beauty and habitability was overlooked, and no reference has been made to the conservation of that resource which alone makes for patriotism, for pure love of country. Our great scenic possessions were not mentioned in the inventory. Nor did the recent meeting of the Conservation Commission, in its programme or in its sessions, consider the question, even though at the moment John Muir was leading a fight to prevent the consummation of a grant by the Secretary of the Interior to San Francisco of a water right that meant the unnecessary giving up of half the Yosemite National Park, and the Secretary of State was

From *The Outlook* 91 (1909), 594–98.

negotiating a Canadian treaty consenting to the diversion, for electric power, of a part of Niagara's flood equal to three-fourths the outflow of Lake Superior. . . .

Now, the present conservation movement looks wholly to the future home of the American people. There is probably enough of everything to last throughout the lifetime of the President who originated and the Governors who have fostered this movement. It is that we may not be justly execrated by our descendants that we now take stock of what remains, after a century of vast advance and vaster waste, and resolve to "conserve the foundations of our prosperity."

This future home of great America—is it to be ugly, unhealthful and in part uninhabitable for an advanced people? Are we to so proceed with the conservation of all our God-given resources but the beauty which has created our love of country, that the generations to come will increasingly spend, in beauty travel to wiser Europe, the millions they have accumulated here, being driven away from what was once a very Eden of loveliness by our careless disregard of appearance? Are our children to grow up in ignorance of the "rocks and rills," the "woods and templed hills," about which they will be trained to sing? Is the national home, the future America, to be designed by us, in this conservation movement, without regard for appearance . . . ?

We are to conserve the soil, the forests, the minerals, and the waters; we can, if we will, and if we plan to that end, at the same time and with complete economy preserve and increase the beauty, healthfulness, and habitability of our country.

How may this be done? By taking thought . . . at the beginning. By considering the appearance of our American home for all nations *while* we "conserve the foundations of our prosperity," and not *after* we have built ugliness, disorder, ill health, and uninhabitability into those foundations. The city of Cleveland is addressing itself to the task of spending millions of money to revise its city plan. Philadelphia has appropriated some two millions to cut across Penn's thoughtlessly wasteful diagonal plan with a parkway. Chicago and St. Louis, Minneapolis and Cincinnati, are planning to put millions into "beauty, healthfulness, and habitability" now where thousands would have sufficed had there been . . . foresight. . . .

Where an American navigable waterway runs through a city, we use it as a sewer, we back our poorest structures upon it, we plant it with glaring billboards. Witness the Milwaukee and the Chicago Rivers, and contrast them with the waterways of Amsterdam and Venice, remembering that the

Venetian canals were primarily designed for commerce by the shrewd Italians who founded that city on an inhospitable marsh. . . .

We are to conserve our minerals. That means we are to dig and mine more carefully, less wastefully. But are we to drag out the earth's interior, with ever so much care, and, with the portions we do not utilize, continue to destroy the attractiveness of the earth's surface? We have probably cared less for economy and for appearance in our coal and ore digging than in the use of any other of the resources we have been wasting. The horrors of the pit-mouth and the mining village are known to but few of us, I believe, else they could not continue in a Christian civilization. The way in which we house the defenseless foreign labor which digs our coal and works our metals, and the naïve wonder we once in a while show at the resulting crime, death, and bad citizenship, are no credit to our sociological intelligence. That it is wasteful and wicked to so cheaply destroy human life we know; but will we Nationally, in this great movement, undertake to make the abstraction of wealth from the earth's bosom less disgustingly ugly, less destructive of human health and comfort and life, than we now do? . . .

We cannot create scenery; we can change it. There will always be scenery, man-made or God-made. It is a question for the National Conservation Commission and its supporters, largely, as to what sort of scenery is to characterize the new America of conservation. Will it be a Niagara of wheel-pits and trail-races and factory wastes, a Yosemite taken for city reservoirs, streams "developed" from beauty to hideousness, waterways through which one prefers to pass in the night, wharves and docks without sightliness or dignity, forests with rectilinear highways regardless of contours or beauty, ore dumps and culm banks continuing to grow, industrial communities with a death rate greater than that of a war? Will we urge expediency rather than actual need as a reason for destroying our great landmarks of natural scenery? Will we more rapidly change the glories of the fairest land the sun ever shone upon to the "conserved" scenery of an unkempt factory yard? . . .

Civic Improvement

MARY RITTER BEARD

THE HUMANITARIAN AND wise planning of beautiful cities and towns is the climax of municipal endeavor, because it represents the coordination of all civic movements looking toward the health, comfort, recreation, education and happiness of urban people.

City planning like all other interests has grown in purpose and scope. From desire for ornamental lampposts has grown a desire for effective light, and not too expensive either. Well-lighted streets become recognized as foes to crime, and out of interest in the lamppost comes an interest in the causes of crime; proper housing, wholesome amusement, and employment may thus be intimately connected with an artistic street lamp.

City planners have not all begun with a lamppost. Some of them began with billboards and thought of billboards exclusively for a long time; then they moved on to municipal art, education, censorship of movies, recreation, housing and labor. Some began with parks and advanced to health and transportation.

There is no one thing in city planning that stands out conspicuously today as the crowning achievement of its purpose. City planning is thus not a finished ideal, but one capable of, and exhibiting, indefinite expansion. In fact, city planning is in its infancy in this country, but its promoters are enthusiasts with a developing sense of values and they are meeting an increasing response among the people for whose interest they are working. . . .

The movement for municipal beauty has been the strongest phase of city planning up to the present time and the element that has appealed to women's civic leagues in their early days very strongly. It is a most legitimate object of civic endeavor and it is comparatively easy of accomplishment where it touches no vital economic interests. "The City Beautiful" only a short time ago was a city with a few wide boulevards, a civic center, handsome

From *Woman's Work in Municipalities* (New York: D. Appleton and Company, 1915), 293–316.

parkways with "Keep Off the Grass" signs in abundance, statues in public squares, public fountains, and public buildings with mural decorations. Alleys and indecent river-front tenements, filthy and narrow side streets, were ignored in the more ostentatious display of mere ornamentation and no provision was made for playgrounds and well-located schools and social centers.

City Planning

The new spirit is rapidly permeating conferences on city planning, however, with an insistence on the elimination of plague spots and unsightly congestion as well as on the creation of boulevards and civic centers. This new spirit is being instilled by women as well as by men. Jane Addams' "The Spirit of Youth and the City Streets" has helped to arouse the feeling that the children are the first to be considered in city plans. Women who have worked for shade trees so extensively have not been unmindful of the fact that mothers have to push baby carriages up and down through the hot sun, oftentimes to the detriment of both mother and child, and they have taught us that mothers should be considered in city plans. In regulating movies women have learned that men are ready to go with their families to a five-cent show in preference to the saloon alone, that the movie has made real inroads upon the saloon, and so they have taught that men should be included in city plans. Thus city planning is becoming of decided human interest and is no longer merely a cultural or artistic recreation. . . .

Women have hailed with pleasure the new slogan "Know Your City," which means that when it is properly known constructive work for improvement will inevitably set in. A good way to know one's city is to have a survey made of it. As we have seen in the chapter on housing reform, women have often organized and made local surveys. In many cities, like Pittsburgh, Scranton, Newburgh, Poughkeepsie, and Cleveland, women helped in working out special features of the surveys. . . .

Controlling Suburbs

Where civic pride and organization promote intelligent efforts in a city to control real estate speculation, unregulated building and congestion, it often happens that the area just outside the city accepts all the evils cast forth by the city. A factory or plant, pushed to the outskirts where a suburb is quickly

developed by land speculators to meet the new housing situation, may easily, and does often, become the center of a community totally without plan
and where the evils of congestion appear in their most exaggerated form.
In some cases, civic leagues of men and women are forming to prevent suburbs coming under such influences, as the city, to which they are neighbor,
agitates for the removal of its factories to the outskirts. . . .

National Vitality, Its Wastes and Conservation

IRVING FISHER

Chapter XIII.—The general value of increased vitality

Section 1.—Disease, poverty, and crime. . . .

We began this report by showing the relation between the conservation of
health and the conservation of wealth. The broadest view of this relation
is, as Emerson has said, that "Health is the first wealth," and as such it is
treated by many economists.

Without enlarging or insisting upon this concept, it is obvious that by
the conservation of health we may ultimately save billions of dollars of wasted
values, and that this conservation is intimately related to conservation of
all other kinds.

We have already seen the vicious circle set up between poverty and disease, each of which tends to produce the other. Metchnikoff contends that
health and morality are correlative, if not interchangeable, terms. A similar idea has been elaborated statistically by Dr. George M. Gould. The
subject is worth much further study. National efficiency is crippled by any
one or all of the parts of the vicious circle—disease, poverty, vice,
vagabondage, crime. It would be interesting to study the tramp problem,
which represents an enormous waste of labor power, in relation to all these
phenomena.

From *Report of the National Conservation Commission*, vol. 3 (1909), 724–49. Original footnotes have been removed.

Section 2.—Conservation of natural resources

It is also true that health begets wealth, and vice versa. Whatever diminishes poverty or increases the physical means of welfare has the improvement of health as one of its first and most evident effects. Therefore an important method of maintaining vital efficiency is to conserve our natural resources—our land, our raw materials, our forests, and our water. Only in this way can we obtain food, clothing, shelter, and the other means of maintaining life. Conversely, the conservation of health will tend in several ways to the conservation of wealth. First of all, the more vigorous and long- lived the race, the better utilization can it make of its natural resources. The labor power of such a race is greater, more intense, more intelligent, and more inventive.

 The development of our natural resources in the future will be more dependent on technical invention than upon the mere abundance of materials.

Just as in warfare it is not so much the gun as the man behind the gun that makes for success, so in industry, as Doctor Shadwell has shown, skill, knowledge, and inventiveness are the chief factors in determining commercial success and supremacy. The backward nations, like China, are characterized by lack of modern inventions. The nations which are industrially most advanced have the railway, the steamship, the power loom, metal working, and innumerable arts and crafts. The change of Japan from a backward to a forward nation is at bottom the introduction of inventions. If conservation prevents lessened fertility, invention makes two blades of grass grow where one grew before.

Future industrial competition will be increasingly a contest of invention. The world rivalry to develop the best system of wireless telegraphy or the best airships is but one example. The future will see the greatest strides taken by the nation which is the most inventive. Now, the primary condition of invention is vitality, a clear brain in a normal body. It is no accident that Edison is a health culturist, or that Krupp, Westinghouse, and other pioneers in industrial development have been men of vigor of mind and body.

Finally, the conservation of health will promote the conservation of other resources by keeping and strengthening the faculty of foresight. One cause of poverty in the individual and the nation is lack of forethought.

One of the first symptoms of racial degeneracy is decay of foresight. Normal, healthy men care for and provide for their descendants. A normal, healthy race of men, and such alone, will enact the laws or develop the public sentiment needed to conserve natural resources for generations yet unborn. When in Rome foresight was lost, care for future generations practically ceased. Physical degeneracy brought with it moral and intellectual degeneracy. Instead of conserving their resources the spendthrift Romans, from the emperor down, began to feed on their colonies and to eat up their capital. Instead of building new structures they used their old coliseum as a quarry and a metal mine.

The problem of the conservation of our natural resources is therefore not a series of independent problems, but a coherent all-embracing whole. If our nation cares to make any provision for its grandchildren and its grandchildren's grandchildren, this provision must include conservation in all its branches—but above all, the conservation of the racial stock itself.

Chapter XIV.—*Things which need to be done*

Section 1. Enumeration of principal measures

In order that American vitality may reach its maximum development, many things need to be done. Among them are the following:

1. The National Government, the States, and the municipalities should steadfastly devote their energies and resources to the protection of the people from disease. Such protection is quite as properly a governmental function as is protection from foreign invasion, from criminals, or from fire. It is both bad policy and bad economy to leave this work mainly to the weak and spasmodic efforts of charity, or to the philanthropy of physicians.
2. The National Government should exercise at least three public health functions: First, investigation; second, the dissemination of information; third, administration.

It should remove the reproach that more pains are now taken to protect the health of farm cattle than of human beings. It should provide more and greater laboratories for research in preventive medicine and public hygiene. Provision should also be made for better and more universal vital statistics, without which it is impossible to know the exact conditions in an epidemic, or, in general, the sanitary or insanitary conditions in any part of the country.

It should aim, as should state and municipal legislation, to procure adequate registration of births, statistics of which are at present lacking throughout the United States.

The National Government should prevent transportation of disease from State to State in the same way as it now provides for foreign quarantine and the protection of the nation from the importation of disease by foreign immigrants. It should provide for the protection of the passenger in interstate railway travel from infection by his fellow-passengers and from insanitary conditions in sleeping cars, etc.

It should enact suitable legislation providing against pollution of interstate streams.

It should provide for the dissemination of information in regard to the prevention of tuberculosis and other diseases, the dangers of impure air, impure foods, impure milk, imperfect sanitation, ventilation, etc. Just as now the Department of Agriculture supplies specific information to the farmer in respect to raising crops or live stock, so should one of the departments, devoted principally to health and education, be able to provide every health officer, school-teacher, employer, physician, and private family with specific information in regard to public, domestic, and personal hygiene.

It should provide for making the national capital into a model sanitary city, free from insanitary tenements and workshops, air pollution, water pollution, food pollution, etc., with a rate of death and a rate of illness among infants and among the population generally so low and so free from epidemics of typhoid or other diseases as will arouse the attention of the entire country and the world. . . .

Conservation By Sanitation:
Air and Water Supply, Disposal of Waste

ELLEN H. RICHARDS

*The spirit in which the problems of modern civilization, especially those in rela-
tion to air supply and ventilation and water supply and waste disposal, are to
be approached and solved.*

THE SANITARY ENGINEER has a treble duty for the next few years
of civic awakening. Having the knowledge, he must be a *leader* in developing works and plants for state and municipal improvement, at the same
time he is an *expert* in their employ. But he must be more; as a health officer
he must be a *teacher* of the people to show them why all these things are
to be. The slowness with which practicable betterments have been adopted
among the rank and file is, partly at least, due to the separation of functions, of specialization, and partly to the exclusiveness of agents in the
work.

The individualism of the nineteenth century extended to the domain of
hygiene. The physician looked after the interest of his patient, not of his
patient's neighbors, and the mass of the people went their own individual
way without giving him a chance until the mischief was done. The advocates of Preventive Medicine, among whom were some of the most eminent physicians, found stony fields for the seed they wished to sow. The
engineer had to plow this field, lift out the stones, and prepare the ground.
The application of sanitary principles used for the benefit of the people with
the same energy and business sense as has been used for the profit of the
individual, will soon prove that sanitation will *pay* as well as railroads and
machine shops. . . .

From *Conservation By Sanitation* (New York: John Wiley & Sons, 1911), pp. v, 60–61, 106–7.

Chapter VI
Economic and Sanitary Efficiency of Waterworks

Does it pay to spend a city's tax money for water of a known and maintained standard? Is it a city's duty? Why not allow the purchase of water as of clothing at risk? For one reason, because the city treasury suffers in an epidemic, and the community suffers many times more, making the cost greater than prevention. But there *may* not be an epidemic. So in the case of a water too hard or too corrosive, why not allow each manufacturer to use whatever rectifier he chooses and not compel him to pay for a municipal water-softening plant? He *may* be a dangerous neighbor if his boiler blows up. It will cost the city more for inspection and for damages.

All this amounts to but one thing. The modern city must pay for assurance against loss of life or of efficiency and consequent financial loss. It might actually take out an insurance policy of a few million dollars against loss by a typhoid-fever epidemic, let us say. This would be good economics, but there is a still better form of assurance. This is protection against epidemics—pure city water for instance. Banks have found the Bankers' Protective Association a decidedly profitable investment. The village, town, or city hesitating between two grades of water must settle the problem by pure economic reasoning. They must figure the risk from the use of each just as the fire-insurance agent would figure his risk on the town hall. The better supply is of course the least risk, but possibly the premium required to install and operate the plant is more than it would be good economy to pay. The premium to be paid in each case must be determined from the reciprocal of the risk rather than from the actual risk itself. That is, the less the risk of infection from a certain source the greater the premium that it is permissible to pay to obtain that source, and vice versa. . . .

Water a national asset.
Conservation of national Resources a public duty.

Preservation of *quality* of water for man's uses is a public duty.

Clean water implies a clean soil. A clean soil can be found only where wastes are properly cared for and the principles and methods of disposal are both understood and faithfully carried out.

Modern civilization demands a close interrelation of all the practical applications of science so that one man's benefit shall not be the detriment of many. It is because of this growing need for wide surveys that government control for the benefit of all the people is becoming more and more thought possible.

The very life of the nation depends on its water supply. As has been stated, food, power, manufacturing, sanitation, as well as personal use, are all intimately bound up with water resources. A study of contamination and purification, of prevention and reclamation, is imperative.

The time is here, already come, when the preservation of the quality and quantity of such water as remains to us is of paramount importance—not only its storage and metered use, but care of the collecting grounds, where the soil must be kept clean because there are no longer vast areas of unused collecting grounds. Fifty years ago both chemists and laymen had a partial justification for their opposition to cremation*; to-day there is little excuse for the continued fouling of the soil. Sanitation means not only clean water but also clean air and clean soil. Clean water is especially dependent on clean soil, and with the great traveling propensity of moderns the rules for clean soil become more and more imperative.

Education by sanitary legislation is being widely considered but is not yet accomplished. Since a goodly part of the duty of *governments* is to educate the people in means for the promotion of their own well-being as well as to make laws, the new social consciousness expects the knowledge gained in the laboratory to be put at the service of the people. In fact legal restrictions are now almost always explained. . . .

* Here cremation refers to the burning of garbage at very high temperatures, not the burning of corpses.—Ed.

Part 4

SMOKE AND CONSERVATION

IN THE CITY

Conservation rhetoric gained remarkable strength in the 1910s, particularly as the American faith in science and government surged in an efficiency craze. The following documents track the changing nature of a specific movement, the effort to control coal smoke in American cities. Gradually conservation rhetoric and practice replaced a less expert-centered and more broadly focused environmental movement to clear urban air.

In the first document, prominent Cincinnati physician and reformer, Charles A. L. Reed, offers an articulate expression of the smoke problem while organizing a Smoke Abatement League in that city. In 1905, Reed addressed the Woman's Club, an organization of middle-class and wealthy women who would help lead the antismoke movement, making a speech that would be reprinted and referred to widely as the movement spread around the country. Reed's speech represents the early antismoke movement, centered in lay activism, particularly among middle-class women and physicians. In the second piece, from 1912, Mrs. Ernest R. Kroeger, President of the St. Louis Women's Organization for Smoke Abatement, recounts the early activism of women in that city's movement, offering a summary of how the early phases of antismoke movements commonly formed around the country.

Even as Kroeger wrote, however, women had begun to lose influence in the antismoke movement, with their rhetoric concerning beauty, health, and morality losing ground to engineers and their rhetoric concerning efficiency and waste. By the mid-1910s, engineers dominated the discussion of smoke and the definition of the "smoke problem." In the third document, Herbert M. Wilson, the Engineer in Charge at the U. S. Bureau of Mines, gives both an overview of the problem and his assertions regarding "The Cure for the Smoke Evil." In this excerpted speech before the American Civic Association, Wilson gives a fairly technical presentation, a sign that soon lay reformers would participate less and less in ever more technical discussions regarding air pollution. He also argues that conservation is economical, not ✳ just virtuous. Four years later, Ernest L. Ohle, a mechanical engineer and a member of the Engineer's Club of St. Louis, gave the then dominant engineering perspective on smoke. Over time, the environmentalist rhetoric of citizens' rights to pure air, as articulated here by Reed, gave way to conservationist rhetoric of efficiency and waste so dominant in Ohle's speech.

An Address on the Smoke Problem

CHARLES A. L. REED

THE SMOKE PROBLEM is in no sense local, but, on the contrary, presents questions of distinctly national interests. It is, in fact, a practical problem that presses for solution upon every class of people in every manufacturing city in the country. But it is a singularly pertinent theme for serious consideration by woman's clubs, if for no other reason than that women, over and above all others, are martyrs to this existing and unfortunately growing order of things. Their rights never seem to be considered by the manufacturer of that class who fancies that in the assumed interest of his business he has a right to manufacture smoke without let or hindrance. The extra drudgery in house-keeping thus imposed upon women is never taken into account by the company whose factories fill the air with soot that filters alike into the parlor and bed-room. The health of women and of children is not a factor in the calculations of the corporation whose power plants load the atmosphere with irritating and otherwise deleterious vapors. A woman's cherished privilege to wear clean clothing, to say nothing of light becoming gowns, and to keep them clean is not regarded by the municipality that permits its streets to become defiled with the grime of its furnaces. For these and for other equally pertinent reasons it is but natural, indeed it is high time that women individually and through their organizations should voice their sentiment by entering a vigorous protest against conditions that, as applied to them, not only violate their sense of decency but that outrage everybody's convictions of justice.

Interests Damaged by Smoke

But martyrs as are women to the smoke nuisance, there are other interests that are equally violated by its existence and perpetuation. Thus it would

Delivered before the Woman's Club of Cincinnati, April 24, 1905. Pamphlet, Cincinnati Historical Society.

be interesting to know if it were possible to ascertain how many thousands of dollars worth of merchandise is annually lost by our dry goods merchants, solely through the ravages of smoke and soot. Clothiers, milliners, dressmakers, tailors, outfitters, grocers, druggists are singularly subject to damage from the same cause. Jewelers are put to extra labor and expense to protect their wares, especially silverplate against the influence of corroding gases that impregnate the atmosphere as the result of imperfect combustion in numerous manufacturing establishments. The damage that has been done and is being done to residence property in Cincinnati and other cities similarly enshrouded with smoke is beyond computation. . . .

Sanitary and Moral Aspects of the Question

The material, that is to say the dollar-and-cents side of the question—and I have alluded to only one part of even that side of the question—is not, however, its only important phase. The sanitary, moral, aesthetic and ethical aspects of the subject have possibly a still further reaching influence— an influence which if reduced to its final analysis, if carried to its ultimate calculation may tell even more seriously on the wrong side of the ledger. "The Cincinnati lung," black and pigmented in contrast with the lungs of those who live and die in the country, was a proverb among physicians until other cities became as smoky and thus deprived Cincinnati of the questionable distinction. The slight morning cough with the equally slight expectoration of black mucous is an experience familiar to the denizens of smoky towns but an experience which, to the medical mind, suggests a persistent although slight irritation of the upper air passages that are thus made hospitable avenues for tuberculous infection. Physicians of smoke-ridden cities testify to the greater frequency there of catarrhal or other disorders of the upper air passages induced by the irritating products of incomplete combustion. But the trouble is not alone physical. It is not, on the face of it, a good thing for any community to become too tolerant of dirt. Physical dirt is close akin to moral dirt and both combined lead to degeneracy. It is precisely against these influences, against this combination of influences, that practical philanthropy is to-day directing its most strenuous efforts in the crowded centers of London, New York, Chicago, and other large cities. It is too much to expect the best results from public schools that exist beneath the somber shade of smoke. A dingy atmosphere is not conducive to a clear intellect. It is difficult to imbue the young with a sense of the beautiful when the beauty itself is bedaubed with soot. It is likewise difficult to instill a sense of justice in the minds of youths who are brought up in a community that permits one interest needlessly but flagrantly and with impunity to violate the equal rights of other interests. Ministers of the gospel would find it an easier task to teach the religion of a clean life and a happy eternity in a material atmosphere less suggestive of a gloomy present and a cheerless hereafter.

The Ethics of the Air

Then, too, there is something to be said about the ethics of the air. Air is necessary to existence. This being true, to breathe pure air must be reckoned among man's inalienable rights. No man has any more right to contaminate the air we breathe than he has to defile the water we drink. No man has any more moral right to throw soot into our parlors than he has to dump ashes into our bed-rooms. No man has any more right to vitiate the air that sustains us than he has to adulterate the food that nourishes us. Poison taken into the body through the lungs is just as much a poison as is some other poison swallowed into the stomach. Poisonous air is probably more disastrous to infants than is adulterated milk. A man's proprietorship extends as distinctly into the air above him as into the earth beneath him. If every man is entitled to the ground he stands upon so is he entitled to the air that envelops him. . . .

Necessity for a National Movement

It is equally impossible for one city, such for instance as St. Paul, or New York or Philadelphia, to change conditions that are essentially national in their extent and influence. One city is afraid to move in the matter for fear that it will drive away some industry to some rival city. This is shown by the letter of a health commissioner of one of our largest interior cities, who, among other things says: "Our attitude towards industrial enterprises of the manufacturing kind has been one of encouragement, and we have been reluctant to deal seemingly harshly with them, and I fancy that this attitude has something to do with the non-enforcement of the smoke ordinances."

It is, therefore, important for this movement, if it is to move, that there should be co-operation between different cities looking to the regulation of the evil. For this reason, it seems that co-operation, to be effective, must be national in extent. . . .

Smoke Abatement in St. Louis

MRS. ERNEST R. KROEGER

THE SMOKE NUISANCE in St. Louis had grown almost intolerable when the Wednesday Club, a strong, fine organization of five hundred women, took up the question and cast about to see what could be done. This was in December, 1910. Up to that time there had been sporadic attempts, with considerable results from these efforts, made by the Smoke Abatement Committee of the Civic League. The Missouri State Law, a strong law covering all aspects of the question excepting that of locomotives, was passed as a result of their work.

The State Law was excellent, but the work of the Civic League in enforcing the law was almost completely hampered by an ineffective City Smoke Abatement Department and indifference on the part of the public. The City Department had combined the Smoke Abatement Department with the Boiler and Elevator Department, and placed at the head of both a Chief Inspector of Boilers and Elevators, with several deputy boiler inspectors, and *no* deputy smoke inspector. The consequence was that there was no force to look after the smoke nuisance.

The Wednesday Club made tentative inquiry of the Civic League as to the necessity of energetic effort, and received in reply a cordial invitation to cooperate with its Smoke Abatement Committee to secure the enforcement of the (existing) smoke ordinance. After accepting the invitation of the Civic League, the Wednesday Club realized that the movement should be larger and more general than a club movement, and, further, felt the necessity of arousing public opinion. With this end in view a mass meeting of the women of St. Louis was called in the Auditorium of the Wednesday Club and a program provided touching on the smoke nuisance from the standpoint of health, cleanliness, housekeeping, city planning, etc. The program included men and women speakers, some of whom were city officials.

From *The American City* 6 (1912), 907–9.

At this meeting, which was crowded, the Women's Organization for Smoke Abatement in St. Louis was formed with 250 paid members. By the next afternoon there were 400 members, and at the present time the membership numbers 1,300. An executive board of twelve women was elected and has had charge of the planning and directing of all the organization's work. These women met weekly the first season and fortnightly the second, and have been enthusiastic and tireless in their crusade against the smoke nuisance.

The first work they took up was districting the city in districts of about five square blocks with volunteer members of the organization as reporters of the smoking chimneys in their districts. Colonel James Gay Butler, one of St. Louis' most public-spirited citizens, came to the assistance of the women with an open purse, stating that he would spend $50,000 if necessary to make St. Louis a clean city. He employed a lawyer and six smoke inspectors to supplement the work of the city, and offered to cooperate with the women in securing legal evidence from their district reports.

These district reports were mailed to the Executive Board of the Women's Organization, where copies were made and forwarded to the newspapers, Colonel Butler's lawyer and the City Department. These cases were then followed up, taken into court and required to comply with the law. For fifteen months Colonel Butler's lawyer and inspectors have secured convictions against offending chimneys, until now the manufacturing districts are pretty well cleaned up. The locomotives, residences and small apartment houses are at present our greatest offenders, and the combined efforts of the women, the Civic League, Colonel Butler and the newspapers are being directed against them. The newspapers have been most powerful allies in the smoke work and have given thousands of dollars of free advertising to the campaign. . . .

At the present writing, the spirit of cooperation extends to all classes of interests in St. Louis. With the Women's Organization, the Civic League, Colonel Butler's force, the press, the City Department, the Business Men's League, the Million Club, and public opinion endorsing a strong adequate ordinance, there is a pretty good chance that it will go into effect, with the result that in a few more years St. Louis will be a clean, beautiful city.

The Cure for the Smoke Evil

HERBERT M. WILSON

AND NOW TO consider the mechanical means by which smoke production may be abated. In its investigations the Geological Survey has found that the chief waste in coal is due to imperfect combustion in furnaces and fire boxes. Steam engines utilize on the average about 8 per cent of the thermal energy of the coal; internal combustion engines utilize less than 20 per cent; and in electric lighting less than 1 per cent of the total energy is rendered available. Perfect combustion means not only a better and more economical utilization of coal but also smokeless conditions. The government is therefore indirectly attempting to abate the smoke nuisance of the country by directly finding how to increase the efficiency with which the coals are used, and thus prolonging or conserving the supply for the nation. The experiments have proven highly successful, so much so that the statement is made by government engineers that the smoke nuisance of American cities is needless and should not be tolerated. Smoke prevention is not only possible, but we stand ready to prove it by actual demonstration at the government experiment station in Pittsburgh. In that smoky city we are producing 1,000 horsepower without smoke, and we are burning under steam boilers coal considered refuse by the trade, costing, delivered at the station, 88 cents a ton. Furthermore, the men detailed to this investigation have found more than 200 plants in the larger cities of Illinois, Indiana, Kentucky, Maryland, Michigan, Missouri, New York, Ohio, and Pennsylvania, which are being operated without smoke and with a gain in economy—for smoke these days means waste. . . .

It is a generally conceded fact that intelligent men trained in boiler-room practice could, by the smokeless combustion of coal, save 10 per cent of the fuel used in 50 per cent of the power and heating plants of the United States, and that in another 25 per cent of the plants such men could save 5 per cent

From *The American City* 4 (1911), 263–67.

of the fuel. It is the practice of nearly all large power plants to employ a boiler-room expert, and many of them have chemists who make frequent tests and investigations to determine the conditions favorable to the best economy. The saving of only a few per cent on the coal consumed will make a handsome return for the cost of the experimental work. There are now competent engineers who make a specialty of supervising boiler plants for a number of firms. . . .

Altogether investigations show that the smokeless American city is entirely possible, and that it will come when the public conscience has thoroughly awakened to the enormous waste of natural and human resources through this evil. The battle being waged by health officers, smoke inspectors and public-spirited citizens in the various cities is not in vain. Through their efforts there has been an awakening of public sentiment, which is to grow into a mightier force as the situation becomes clearer to the people. It is true that a large number of fuel users in the United States still believe that smoke means wealth, and that to produce less smoke would add to the cost of running their mills and factories; but these individuals are becoming fewer in number each day as the fallacy of their views is demonstrated. The general body of fuel users is still permeated with the idea that it can not suppress black smoke without increasing wages or fuel bills, whereas the reverse is the case. With lamentable indifference to the health of the community obliged to live within range of their factory chimneys they continue to burn coal on the unscientific rule of thumb method handed down from the last generation. Nevertheless the truth is gaining ground. The more intelligent manufacturers see the economy in the smokeless combustion of coal, and when the argument touches the pocketbook it is bound to spread. . . .

Altogether the situation is hopeful. The smokeless city in the future is to be the note of civilization; a smoky city is to be the sign of barbarism, and not the badge of prosperity some have boasted it. The few agitators for emancipation from the evil of soot of a few years ago have been reinforced by a vast army of crusaders. Black, dust-laden smoke has been proved to be wasteful. It is economy to have smokeless mills, factories and cities, and the converts this consideration is daily making to the reform, bid fair to rid us soon of the spreading, insidious, heaven-obscuring nuisance.

Smoke Abatement: A Report on Recent Investigations Made at Washington University

ERNEST L. OHLE

IT IS SCARCELY necessary before a body of engineers to go into the causes for the pall of smoke that hangs almost continually over our city. Nor is it necessary to prove to a St. Louisan that smoke is a nuisance.

Our greatest fault lies in the fact that we too frequently forget that smoke can be abated and that the abatement may be, and nearly always is, a source of profit to the smoke maker as well as to the community.

Between 8,000,000 and 9,000,000 tons of soft coal, according to the Merchants' Exchange, are received annually in St. Louis, and the greater part of this is burned in the St. Louis district. The average efficiency with which this coal is burned according to the best estimates, is probably not far from 50 per cent. With proper installations and operation there should be an efficiency of 60 per cent or a saving of 20 per cent over present conditions. This means that we are practically throwing away between 1,500,000 and 2,000,000 tons of coal a year, to say nothing of the expense of handling this coal and the ashes formed from it. Add to this waste the cost of removing the effects of the smoke and you have an enormous sum which goes into the production of nothing of economic value.

The common oil lamp is one of the best illustrations of the perfect combustion and consequent smoke prevention. The heated gases rising in the chimney produce a draft and fresh air is continually drawn in at the bottom through the hot gauze, which warms and divides it so as to insure thorough mixing with the gases from the burning fuel. Turn up the wick and the flame becomes smoky—too much hydrocarbon for the air supply; raise the chimney slightly from the bottom and again there is smoke—too much air at too low a temperature which chills the flame; insert a cool metal rod into the chimney and soot is deposited on it— chilling of the flame again and disengaging of the carbon, while the hydrogen continues to burn. Thus we

From *Association of Engineering Societies* 55 (Nov. 1915), 139–45.

see that there are three conditions necessary for perfect combustion—sufficient air, thorough mixing of the gases and a sustained high temperature. If these conditions are present in a boiler plant there will be nothing to fear from the smoke inspector.

Many of the existing plants are not equipped to give these conditions—though much more could be accomplished than is being done. A recent issue of the *Engineering Magazine* characterized as Stupidity and Waste the losses which appear in the power plant, and there is no place where a greater saving can be effected than in the boiler room. It is amazing how little information the average plant owner has in regard to the costs in his boiler room. According to one authority one fourth of the fuel bill is controlled by the fireman—how is that fourth being managed? Only by accurate records of each day's operation can any idea at all be gained as to its disposition.

According to one of the largest coal dealers in St. Louis he was furnishing five tons of coal a day to a certain plant—the operation of this plant was taken over by another company and the fuel consumption immediately dropped to four tons a day using the same fuel and under the same load conditions. One of the most notable savings was that of the Crucible Steel Company of America. Their boiler plant was reconstructed at a cost of $130,000, and a saving of $60,000 was effected the first year. . . .

In our investigations last spring a large number of power plants were visited in company with an inspector from the City department and smoke observations made. The installations that were examined included fire and water-tube boiler plants and the furnaces ranged through hand fired of the plain grate and down draft types and automatic stokers of the chain grate, inclined grate, and underfeed types.

It was found that some of the best equipped plants were among the worst offenders, due to careless operation and wrong sized coal, and that some of the poorly equipped plants were giving practically no trouble at all, due to careful operation or to the simplest of devices. . . .

Part 5

CONSERVATION, PRESERVATION,

AND HETCH HETCHY

Perhaps the most famous story in American environmental history, the battle for Hetch Hetchy punctuates the Progressive Era conservation movement. Here the city of San Francisco adopted conservation rhetoric in its effort to acquire the right to dam the Tuolumne River inside Yosemite National Park. The plan even gained the support of national conservation figures, including Gifford Pinchot. On the other side of the debate, preservationists, led by Sierra Club founder John Muir, argued against the plan and against the rhetoric of consumption that surrounded the conservationist effort. The Hetch Hetchy story has traditionally been depicted as a moment of terrible conflict between "utilitarian" and "preservationist" wings of the conservation movement—revealing one of the many rifts in the environmental movement that persist to this day.

In the first document, Warren Olney presents the argument in favor of the proposed dam in Hetch Hetchy. Olney, a founding member of the Sierra Club in 1892, had been a longtime friend of John Muir and had served as one of the club's directors. By the time of the Hetch Hetchy debate, Olney was the mayor of Oakland, and his conflict with other club members forced him to resign his membership and it cost him his friendship with Muir. In an article appearing in the same issue of *Out West*, Edward Taylor Parsons argues against the dam, emphasizing the many options San Francisco had. Like Olney, Parsons served on the Sierra Club Board of Directors; he was also among the club's most active mountaineers. Indeed, the club had been formed around the love of such outings in the Sierra Nevada, and gradually became as much a political organization to protect the mountains as an outings club to promote their use. Like many of the club's early members, Parsons was a San Francisco Bay area businessman, eager to protect the great natural resources near his adopted city.

The third document comes from John Muir, the man who led the fight against the Hetch Hetchy plan, and the most important figure in the Sierra Club. Even before the Hetch Hetchy controversy grew into a national debate, Muir had become well known in the United States through his nature writings, including the very popular book *Our National Parks,* published in 1901. Working with the help of Robert Underwood Johnson, the publisher of *Century* magazine (and the book excerpted here), Muir publicized the beauty of Hetch Hetchy specifically and the value of wilderness generally. In one of the most famous passages in the history of American environmentalism, Muir attacked the dam plan, and conservationism, with religious zeal.

In late 1913, Muir and the other opponents to the dam plan lost the fight when President Woodrow Wilson signed the bill giving San Francisco the right to build within Yosemite National Park. Ironically, both Parsons and Muir died the next year.

Water Supply for the Cities
About the Bay of San Francisco

WARREN OLNEY

THE CITY OF San Francisco has now a population approaching five hundred thousand. The cities of Richmond, Berkeley, Oakland, Alameda, San Jose, and the towns and villages between, all bordering on the Bay, aggregate as many more. This population is bound to increase rapidly in numbers. How great the population will be thirty-three years from now, which we will assume to be the average life of a generation, no man can tell. But it will certainly be counted by the millions.

The streams flowing into the Bay of San Francisco are barely sufficient to supply the present population with water, but are not sufficient for the needs of the near future, nor are they sufficient for the present if we should have two or three successive years of drought. As civilization advances, the consumption of water per capita rapidly increases, so that if you add the increasing use of water per capita to the rapidly increasing population, the cities and communities about the Bay of San Francisco will shortly be without a sufficient supply, unless water is obtained from some other source than the immediate neighborhood.

It is useless to talk about increasing the output of potable water from the streams about the Bay. The quantity can undoubtedly be increased from the Alameda water-shed, but if all of the water obtainable from that source is properly utilized, the cities about the Bay must still, in the near future, go to the Sierra Nevada for their water. The water supply from the streams around the Bay is owned by corporations and individuals, who put an extravagant estimate upon the value of their water plants and resources. San Francisco is supplied by the Spring Valley Water Works, Oakland and Berkeley, etc., by the People's Water Company, and both of those corporations maintain, and will always maintain, in the courts and everywhere else, that their property rights are greatly enhanced in value by there being no other avail-

From *Out West* 31 (July 1909), 599–605.

able source of water supply for the people in the vicinity of San Francisco. All experience shows that a city should own its own water works, and own or control its sources of water supply. The opinion has become almost universal among the people about the Bay Cities that the plant and water supply of the two corporations above named should be purchased, if they can be obtained at anything like a reasonable figure, because these corporations already possess quite complete distributing systems, and also because they or their predecessors have been furnishing water ever since there was a demand for it. This opinion is based upon sound reasoning and is backed up by experience of other cities and by ethical and economic laws.

So here is the situation that confronts the people about the Bay of San Francisco: Their sources of water supply belong to private corporations, or public utility corporations if that term is preferred, and that supply will be inadequate in the very near future. They must purchase, and desire and intend to purchase, the local water plants and water resources, but to meet the needs of the increasing population and the growing needs of civilization, more water must be obtained from a distance. An abundance of water can only be obtained from a few of the streams on the western slope of the Sierra Nevada Mountains, which streams flow into the Sacramento and San Joaquin Rivers. All of the waters of those streams, except one, have already been appropriated by water-power companies, irrigation companies and mining companies. Except at a frightful cost, a cost entirely beyond the present capacity of the people to pay, the water from these streams cannot be obtained. But there is one stream that does have a sufficient flow of water to satisfy the needs of all of the cities around the Bay of San Francisco for many generations to come, that is free from prior claims or locations of all private corporations, so far as its flowing waters are concerned, except the claims of two irrigation districts. . . .

This river is the Tuolumne River. It has the largest water-shed and the best water-shed, and has a larger flow, and far and away a better reservoir site, than any of the other streams accessible to the people. Shall the people be refused the use of this water and compelled either to go without Sierra water or to assess themselves ten or twenty millions of dollars more to get water from somewhere else? Is it possible that any intelligent lover of his race can answer the above question in the affirmative?

There have been many objections made to San Francisco and the other Bay Cities utilizing the waters of Tuolumne River, but those objections have been made mostly by people who do not understand the actual condition that confronts the Bay Cities, or else are lacking in the ability to take a broad, comprehensive view of the situation and the needs of humanity. I will take up the objections made *seriatim,* but before doing so let me call attention to the attitude of President Roosevelt, of Secretary James A. Garfield, and of Mr. Gifford Pinchot, all three of whom are as enthusiastic lovers of nature as any that can be found, and who have done more for conservation for the use of all the people of the natural resources of the country than any other three men in our history. There are no greater enthusiasts for nature and the beauties of nature than these three men. Now what has been their attitude in regard to the City of San Francisco utilizing the waters of the Tuolumne River? President Roosevelt strongly favored the plan. Secretary Garfield was its most earnest advocate before the committees of Congress, and Mr. Gifford Pinchot added the great weight of his experience and love of the woods and mountains to the arguments urged by the people of the Bay Cities. In my opinion, Secretary Garfield made the best argument before the committee of the House of Representatives, urging the passage of the bill to which I shall presently refer, that was made by anyone.

I attended and took part in some of the hearings before that committee, and became convinced, when the Spring Valley Water Works opposed the bill, that a fellow-feeling on the part of other interests and a desire to hit the Roosevelt administration had a hundred times more influence than did the arguments of some of the so-called "nature lovers" who opposed what was desired by the Bay Cities on the ground that it would tend to destroy the natural beauties of Hetch Hetchy Valley and injure the country above the valley as a place of resort for nature lovers in the summer months. That is to say, the arguments of the so-called nature lovers had really very little influence, but the sympathy between financial interests desiring to use national resources for personal exploitation and to get even with the Administration had everything to do with it. The nature lovers merely furnished to these enemies of the public good arguments and excuses for not granting to the Bay Cities what they so much desire and so greatly need.

But as citizens of California have furnished the arguments for the "Interests" and have stirred up many people in the East who are not acquainted with the situation and are influenced by sentiment without regard to the actual necessities of the people, I will now give a little attention to the points made by them. . . .

The first ground of opposition urged was that the erection of a dam at the lower end of Hetch Hetchy Valley, thereby creating an artificial lake covering the entire floor of the valley, would destroy the attractions of a most beautiful and interesting mountain valley. . . .

To make a lake of this valley of course will destroy the meadow, but the lake that will be created will be a much greater natural attraction than the valley in its present condition. The lower end of the valley is a wet meadow, and the mosquitoes constitute a frightful pest. In ordinary seasons it is not until late in July that people can camp in the valley with comfort. Very few people visit the valley. Its character has been known for more than forty years. I spent eight days in the valley last summer, after the mosquito season had passed, and I do not believe twenty persons altogether, besides United States soldiers, were there during the time I was. If the recommendations of President Roosevelt, Secretary Garfield and Mr. Pinchot are adopted, San Francisco will turn this beautiful but mosquito-breeding meadow into a beautiful mountain lake, whose attractions will be unique in character and probably as great as those of any lake of its size in the mountains of any country. . . .

The charms of Hetch Hetchy Valley have been known for more than forty years, but it is rare to find any person in California who has taken the trouble to go to see it. If San Francisco is allowed to turn the valley into a reservoir, she will have to build good roads and make the valley accessible. Then, no doubt, there will be a hundred visitors where there is one now. . . .

Before closing, attention should be called to another matter: Locations subsequent in date to those made by Mayor Phelan for the benefit of San Francisco have been made and filed, and if the Bay Cities are not allowed to impound and use the water of the Tuolumne River, private corporations are preparing to use them for power and for sale. In fact, I suspect that the hostile influences at work to defeat the desires and the reasonable requirements of our people have been in part inspired by these late claimants to Tuolumne water. Who shall have the use of the water flowing in this mountain river? Shall it be the people, millions of whom need it, or private corporations? This water will not be allowed to go to waste. If the Bay Cities do not get it, private corporations certainly will.

Proposed Destruction Of Hetch-Hetchy

E. T. PARSONS

THERE IS BUT one great National Park in California—The Yosemite National Park. The other two National Parks, the Sequoia and General Grant, are small by comparison, and were created to preserve groves of our "big trees." Not only is the Yosemite National Park one of the most important parks in America, but it is unrivaled in the whole world. Yet this incomparable wonderland—this majestic playground belonging to all the people of the nation, is threatened with destructive invasion in order that selfish and local interests may profit in a financial way. The proposed violation of the Yosemite National Park is not only absolutely unnecessary, but it is questionable whether it would, from an economic standpoint, be for the best interest of the community seeking the destructive privilege. If the needless and destructive right to flood the wonderful Hetch-Hetchy Valley is granted to San Francisco, the precedent that would be established would shake to the very foundation the whole National Park policy. Thereafter, no National Park, however great and wonderful, would be safe from despoliation; for this instance would be pointed out as an example where a nation had sanctioned a most destructive trespass upon one of the greatest scenic wonders of the world. But fortunately the people of this nation are rapidly awakening to the seriousness of the danger that threatens one of its most priceless possessions. They are already appealing to Congress to stop the mischief before there is possibility of its being consummated. . . .

One of the proponents of the Hetch-Hetchy water project has been quoted as saying that the Hetch-Hetchy Valley is "a rich man's playground." It indicates how little he knows of his subject, for, from personal knowledge, I can assert that the overwhelming majority of those who have visited the Hetch-Hetchy Valley have been persons to whom even the slight expenditure involved in the trip was a financial sacrifice.

From *Out West* 31 (July 1909), 607–27.

When this imperial State shall have become settled as the voice of Destiny seems to have decreed, and the San Joaquin Valley is teeming with a countless population, those tillers of field and vineyard will look to the mountains as a place of refuge from the great heat of the summer months. The campers in wagons from the plains are already seeking health and recreation in Yosemite Valley in the summer months by the thousand, and it is well known that the place is already becoming crowded to the point of discomfort. The other available places which these tired, hard-working sons of toil will naturally seek are the Big Tuolumne Meadows and Hetch-Hetchy Valley.

Instead of being a "rich man's playground," the Hetch-Hetchy Valley is destined to be primarily a health-giving resort for the wage-earner.

The Hetch-Hetchy Valley has been called "swampy" and a "mosquito-meadow," etc., by the zealous advocates of the city. There is no more certain indication of a losing cause than a resort by its proponents to misrepresentation. I have seen the Merced River so high in flood-time that a large portion of the floor of the Yosemite Valley was converted into a temporary lake. I have experienced attacks of mosquitoes in the Bridal Veil Meadows at the lower end of the Yosemite Valley that would put the Hetch-Hetchy cohorts to shame. Such arguments would be equally applicable to damming Yosemite itself. Only the lower third of the Hetch-Hetchy Valley is subject to temporary flooding, and the mosquitoes there last but a short time each season. The upper two-thirds is a high landscape garden, beautified by exquisite groves of mighty oaks and carpeted with flowers and ferns. As is the case with the Yosemite, a system of drainage and a liberal use of petroleum will eradicate the mosquito nuisance. The advocates of this water system say that the Hetch-Hetchy is inaccessible and can only be visited three months in the year. This is a poor reason for destroying it when it can be made easily accessible with the expenditure of a few thousand dollars and can eventually be kept open to the public throughout the year. These arguments would have been equally applicable to the Yosemite a few years ago.

It is often given as a reason for sacrificing this finest half of the Park, that comparatively few resort to it at the present time. We are not opposing this invasion of our greatest park because of the present. Even should the city succeed in damming Hetch-Hetchy, it could not well do so before most of us would have revisited it many times. We are not actuated by selfish motives, though we have been called "hoggish and mushy esthetes." If it were only our personal pleasure that would be jeopardized, San Francisco could have the Hetch-Hetchy Valley a thousand times over. . . .

Fortunately we are looking further into the future than many who have discussed this subject. Measured by the present, San Francisco has no need for the Hetch-Hetchy Valley. Then why measure the need of this wonderful region for a national park by the present travel? . . .

In order to prejudice in their favor the uninformed, the Hetch-Hetchy advocates have proclaimed that damming the Hetch-Hetchy Valley will enhance its scenery by converting it into a beautiful mountain lake. The first answer to this bit of sophistry is that in the surrounding mountains countless beautiful lakes abound, while there is only one Hetch-Hetchy Valley. Once destroy its floor by flooding, and unequaled camp grounds that will accommodate thousands of persons will be obliterated. The walls are so precipitous that it could not be viewed with ease and comfort except from a very few places and from the artificial scar at the dam site, which is far removed from the more beautiful portions of the Valley. If one could not live and camp on the floor of the Valley and enjoy its wonders at leisure, how many would take the long trip to see a reservoir from a dam? . . .

It is doubtful if any city in the world of her size has more available sources than has San Francisco. Eminent hydraulic engineers have endorsed many of the following sources from which San Francisco can obtain a water supply [Parsons lists fourteen different options, including the American and Sacramento Rivers and Lake Tahoe]. . . .

We do not contend that all of these sources are available and desirable, but many of them are. . . .

Shall this matchless region, where the voice of Nature whispers in softest harmony and anon rises to thunder tones, where countless charms of form and color glisten in the sun, where rugged grandeur and delicate tracery appear in endless panorama to rejoice the eye—shall this be forever closed and barred from the enjoyment of the present and of posterity?

Passing from the Tuolumne cañon into Hetch-Hetchy is like entering a haven of peace after a storm. Here, too, are stern granite cliffs and the sound of falling waters, but here we do not need to live so close under the shadow of the frowning walls nor feel the ground tremble with the cataract's force. Instead, we move through a wonderful garden, shoulders abrush with tall grasses or the yellow blossoms of the evening primrose, through wonderful groves of fir, of pine, of libocedrus, or of giant oaks. Here are spacious, beautiful camping grounds for thousands beside the smoothly flowing river, with vistas through the trees of tall Kolana Dome, of the mighty Hetch-Hetchy Fall, or of delicate Tueeulala. Here is a garden of paradise, shut in from the troubled outside world by blue-creviced cliffs, lurking place of mysterious shadows by day; by night, when the moon shines, a realm of ghostly phantasy, where fairies might weave their fabric of dreams. . . .

Hetch Hetchy Valley

JOHN MUIR

YOSEMITE IS SO wonderful that we are apt to regard it as an exceptional creation, the only valley of its kind in the world; but Nature is not so poor as to have only one of anything. Several other yosemites have been discovered in the Sierra that occupy the same relative positions on the Range and were formed by the same forces in the same kind of granite. One of these, the Hetch Hetchy Valley, is in the Yosemite National Park about twenty miles from Yosemite and is easily accessible to all sorts of travelers by a road and trail that leaves the Big Oak Flat road at Bronson Meadows a few miles below Crane Flat, and to mountaineers by way of Yosemite Creek basin and the head of the middle fork of the Tuolumne.

It is said to have been discovered by Joseph Screech, a hunter, in 1850, a year before the discovery of the great Yosemite. After my first visit to it in the autumn of 1871, I have always called it the "Tuolumne Yosemite," for it is a wonderfully exact counterpart of the Merced Yosemite, not only in its sublime rocks and waterfalls but in the gardens, groves and meadows of its flowery park-like floor. The floor of Yosemite is about 4000 feet above the sea; the Hetch Hetchy floor about 3700 feet. And as the Merced River flows through Yosemite, so does the Tuolumne through Hetch Hetchy. The walls of both are of gray granite, rise abruptly from the floor, are sculptured in the same style and in both every rock is a glacier monument.

From *The Yosemite* (New York: Century, 1912), 249–62.

Standing boldly out from the south wall is a strikingly picturesque rock called by the Indians, Kolana, the outermost of a group 2300 feet high, corresponding with the Cathedral Rocks of Yosemite both in relative position and form. On the opposite side of the Valley, facing Kolana, there is a counterpart of the El Capitan that rises sheer and plain to a height of 1800 feet, and over its massive brow flows a stream which makes the most graceful fall I have ever seen. From the edge of the cliff to the top of an earthquake talus it is perfectly free in the air for a thousand feet before it is broken into cascades among talus boulders. It is in all its glory in June, when the snow is melting fast, but fades and vanishes toward the end of summer. The only fall I know with which it may fairly be compared is the Yosemite Bridal Veil; but it excels even that favorite fall both in height and airy-fairy beauty and behavior. Lowlanders are apt to suppose that mountain streams in their wild career over cliffs lose control of themselves and tumble in a noisy chaos of mist and spray. On the contrary, on no part of their travels are they more harmonious and self-controlled. Imagine yourself in Hetch Hetchy on a sunny day in June, standing waist-deep in grass and flowers (as I have often stood), while the great pines sway dreamily with scarcely perceptible motion. Looking northward across the Valley you see a plain, gray granite cliff rising abruptly out of the gardens and groves to a height of 1800 feet, and in front of it Tueeulala's silvery scarf burning with irised sun-fire. In the first white outburst at the head there is abundance of visible energy, but it is speedily hushed and concealed in divine repose, and its tranquil progress to the base of the cliff is like that of a downy feather in a still room. Now observe the fineness and marvelous distinctness of the various sun-illumined fabrics into which the water is woven; they sift and float from form to form down the face of that grand gray rock in so leisurely and unconfused a manner that you can examine their texture, and patterns and tones of color as you would a piece of embroidery held in the hand. Toward the top of the fall you see groups of booming, comet-like masses, their solid, white heads separate, their tails like combed silk interlacing among delicate gray and purple shadows, ever forming and dissolving, worn out by friction in their rush through the air. Most of these vanish a few hundred feet below the summit, changing to varied forms of cloud-like drapery. Near the bottom the width of the fall has increased from about twenty-five feet to a hundred feet. Here it is composed of yet finer tissues, and is still without a trace of disorder— air, water and sunlight woven into stuff that spirits might wear.

So fine a fall might well seem sufficient to glorify any valley; but here, as in Yosemite, Nature seems in nowise moderate, for a short distance to the eastward of Tueeulala booms and thunders the great Hetch Hetchy Fall, Wapama, so near that you have both of them in full view from the same standpoint. It is the counterpart of the Yosemite Fall, but has a much greater volume of water, is about 1700 feet in height, and appears to be nearly vertical, though considerably inclined, and is dashed into huge outbounding bosses of foam on projecting shelves and knobs. No two falls could be more unlike—Tueeulala out in the open sunshine descending like thistledown; Wapama in a jagged, shadowy gorge roaring and thundering, pounding its way like an earthquake avalanche.

Besides this glorious pair there is a broad, massive fall on the main river a short distance above the head of the Valley. Its position is something like that of the Vernal in Yosemite, and its roar as it plunges into a surging trout-pool may be heard a long way, though it is only about twenty feet high. On Rancheria Creek, a large stream, corresponding in position with the Yosemite Tenaya Creek, there is a chain of cascades joined here and there with swift flashing plumes like the one between the Vernal and Nevada Falls, making magnificent shows as they go their glacier-sculptured way, sliding, leaping, hurrahing, covered with crisp clashing spray made glorious with sifting sunshine. And besides all these a few small streams come over the walls at wide intervals, leaping from ledge to ledge with birdlike song and watering many a hidden cliff-garden and fernery, but they are too unshowy to be noticed in so grand a place.

The correspondence between the Hetch Hetchy walls in their trends, sculpture, physical structure, and general arrangement of the main rock-masses and those of the Yosemite Valley has excited the wondering admiration of every observer. We have seen that the El Capitan and Cathedral rocks occupy the same relative positions in both valleys; so also do their Yosemite points and North Domes. Again, that part of the Yosemite north wall immediately to the east of the Yosemite Fall has two horizontal benches, about 500 and 1500 feet above the floor, timbered with golden-cup oak. Two benches similarly situated and timbered occur on the same relative portion of the Hetch Hetchy north wall, to the east of Wapama Fall, and on no other. The Yosemite is bounded at the head by the great Half Dome. Hetch Hetchy is bounded in the same way, though its head rock is incomparably less wonderful and sublime in form.

The floor of the Valley is about three and a half miles long, and from a fourth to half a mile wide. The lower portion is mostly a level meadow about a mile long, with the trees restricted to the sides and the river banks, and partially separated from the main, upper, forested portion by a low bar of glacier-polished granite across which the river breaks in rapids.

The principal trees are the yellow and sugar pines, digger pine, incense cedar, Douglas spruce, silver fir, the California and golden-cup oaks, balsam cottonwood, Nuttall's flowering dogwood, alder, maple, laurel, tumion, etc. The most abundant and influential are the great yellow or silver pines like those of Yosemite, the tallest over two hundred feet in height, and the oaks assembled in magnificent groves with massive rugged trunks four to six feet in diameter, and broad, shady, wide-spreading heads. The shrubs forming conspicuous flowery clumps and tangles are manzanita, azalea, spiraea, brier-rose, several species of ceanothus, calycanthus, philadelphus, wild cherry, etc.; with abundance of showy and fragrant herbaceous plants growing about them or out in the open in beds by themselves—lilies, Mariposa tulips, brodiaeas, orchids, iris, spraguea, draperia, collomia, collinsia, castilleja, nemophila, larkspur, columbine, goldenrods, sunflowers, mints of many species, honeysuckle, etc. Many fine ferns dwell here also, especially the beautiful and interesting rock-ferns—pellaea, and cheilanthes of several species—fringing and rosetting dry rock-piles and ledges; woodwardia and asplenium on damp spots with fronds six or seven feet high; the delicate maidenhair in mossy nooks by the falls, and the sturdy, broad-shouldered pteris covering nearly all the dry ground beneath the oaks and pines.

It appears, therefore, that Hetch Hetchy Valley, far from being a plain, common, rock-bound meadow, as many who have not seen it seem to suppose, is a grand landscape garden, one of Nature's rarest and most precious mountain temples. As in Yosemite, the sublime rocks of its walls seem to glow with life, whether leaning back in repose or standing erect in thoughtful attitudes, giving welcome to storms and calms alike, their brows in the sky, their feet set in the groves and gay flowery meadows, while birds, bees, and butterflies help the river and waterfalls to stir all the air into music—things frail and fleeting and types of permanence meeting here and blending, just as they do in Yosemite, to draw her lovers into close and confiding communion with her.

Sad to say, this most precious and sublime feature of the Yosemite National Park, one of the greatest of all our natural resources for the uplifting joy and peace and health of the people, is in danger of being dammed and made into a reservoir to help supply San Francisco with water and light, thus flooding it from wall to wall and burying its gardens and groves one or two hundred feet deep. This grossly destructive commercial scheme has long been planned and urged (though water as pure and abundant can be got from sources outside of the people's park, in a dozen different places), because of the comparative cheapness of the dam and of the territory which it is sought to divert from the great uses to which it was dedicated in the Act of 1890 establishing the Yosemite National Park.

The making of gardens and parks goes on with civilization all over the world, and they increase both in size and number as their value is recognized. Everybody needs beauty as well as bread, places to play in and pray in, where Nature may heal and cheer and give strength to body and soul alike. This natural beauty-hunger is made manifest in the little window-sill gardens of the poor, though perhaps only a geranium slip in a broken cup, as well as in the carefully tended rose and lily gardens of the rich, the thousands of spacious city parks and botanical gardens, and in our magnificent National parks—the Yellowstone, Yosemite, Sequoia, etc.—Nature's sublime wonderlands, the admiration and joy of the world. Nevertheless, like anything else worth while, from the very beginning, however well guarded, they have always been subject to attack by despoiling gain-seekers and mischief-makers of every degree from Satan to Senators, eagerly trying to make everything immediately and selfishly commercial, with schemes disguised in smug-smiling philanthropy, industriously, shampiously crying, "Conservation, conservation, panutilization," that man and beast may be fed and the dear Nation made great. Thus long ago a few enterprising merchants utilized the Jerusalem temple as a place of business instead of a place of prayer, changing money, buying and selling cattle and sheep and doves; and earlier still, the first forest reservation, including only one tree, was likewise despoiled. Ever since the establishment of the Yosemite National Park, strife has been going on around its borders and I suppose this will go on as part of the universal battle between right and wrong, however much its boundaries may be shorn, or its wild beauty destroyed.

The first application to the Government by the San Francisco Supervisors for the commercial use of Lake Eleanor and the Hetch Hetchy Valley was made in 1903, and on December 22nd of that year it was denied by the Secretary of the Interior, Mr. Hitchcock, who truthfully said:

> Presumably the Yosemite National Park was created such by law because of the natural objects of varying degrees of scenic importance located within its boundaries, inclusive alike of its beautiful small lakes, like Eleanor, and

its majestic wonders, like Hetch Hetchy and Yosemite Valley. It is the aggregation of such natural scenic features that makes the Yosemite Park a wonderland which the Congress of the United States sought by law to reserve for all coming time as nearly as practicable in the condition fashioned by the hand of the Creator—a worthy object of National pride and a source of healthful pleasure and rest for the thousands of people who may annually sojourn there during the heated months.

In 1907 when Mr. Garfield became Secretary of the Interior the application was renewed and granted; but under his successor, Mr. Fisher, the matter has been referred to a Commission, which as this volume goes to press still has it under consideration.

The most delightful and wonderful camp grounds in the Park are its three great valleys—Yosemite, Hetch Hetchy, and Upper Tuolumne; and they are also the most important places with reference to their positions relative to the other great features—the Merced and Tuolumne Cañons, and the High Sierra peaks and glaciers, etc., at the head of the rivers. The main part of the Tuolumne Valley is a spacious flowery lawn four or five miles long, surrounded by magnificent snowy mountains, slightly separated from other beautiful meadows, which together make a series about twelve miles in length, the highest reaching to the feet of Mount Dana, Mount Gibbs, Mount Lyell and Mount McClure. It is about 8500 feet above the sea, and forms the grand central High Sierra camp ground from which excursions are made to the noble mountains, domes, glaciers, etc.; across the Range to the Mono Lake and volcanoes and down the Tuolumne Cañon to Hetch Hetchy. Should Hetch Hetchy be submerged for a reservoir, as proposed, not only would it be utterly destroyed, but the sublime cañon way to the heart of the High Sierra would be hopelessly blocked and the great camping ground, as the watershed of a city drinking system, virtually would be closed to the public. So far as I have learned, few of all the thousands who have seen the park and seek rest and peace in it are in favor of this outrageous scheme.

One of my later visits to the Valley was made in the autumn of 1907 with the late William Keith, the artist. The leaf-colors were then ripe, and the great godlike rocks in repose seemed to glow with life. The artist, under their spell, wandered day after day along the river and through the groves and gardens, studying the wonderful scenery; and, after making about forty sketches, declared with enthusiasm that although its walls were less sublime in height, in picturesque beauty and charm Hetch Hetchy surpassed even Yosemite.

That any one would try to destroy such a place seems incredible; but sad experience shows that there are people good enough and bad enough for anything. The proponents of the dam scheme bring forward a lot of bad arguments to prove that the only righteous thing to do with the people's parks is to destroy them bit by bit as they are able. Their arguments are curiously like those of the devil, devised for the destruction of the first garden— so much of the very best Eden fruit going to waste; so much of the best Tuolumne water and Tuolumne scenery going to waste. Few of their statements are even partly true, and all are misleading.

Thus, Hetch Hetchy, they say, is a "low-lying meadow." On the contrary, it is a high-lying natural landscape garden, as the photographic illustrations show.*

"It is a common minor feature, like thousands of others." On the contrary it is a very uncommon feature; after Yosemite, the rarest and in many ways the most important in the National Park.

"Damning and submerging it 175 feet deep would enhance the beauty by forming a crystal-clear lake." Landscape gardens, places of recreation and worship, are never made beautiful by destroying and burying them. The beautiful sham lake, forsooth, would be only an eyesore, a dismal blot on the landscape, like many others to be seen in the Sierra. For, instead of keeping it at the same level all the year, allowing Nature centuries of time to make new shores, it would, or course, be full only a month or two in the spring, when the snow is melting fast; then it would be gradually drained, exposing the slimy sides of the basin and shallower parts of the bottom, with the gathered drift and waste, death and decay of the upper basins, caught here instead of being swept on to decent natural burial along the banks of the river or in the sea. Thus the Hetch Hetchy dam-lake would be only a rough imitation of a natural lake for a few of the spring months, and an open sepulcher for the others.

"Hetch Hetchy water is the purest of all to be found in the Sierra, unpolluted, and forever unpollutable." On the contrary, excepting that of the Merced below Yosemite, it is less pure than that of most of the other Sierra streams, because of the sewerage of camp grounds draining into it, especially of the Big Tuolumne Meadows camp ground, occupied by hundreds

---

* Two photographs of Hetch Hetchy appeared in this chapter.—Ed.

of tourists and mountaineers, with their animals, for months every summer, soon to be followed by thousands from all the world.

These temple destroyers, devotees of ravaging commercialism, seem to have a perfect contempt for Nature, and, instead of lifting their eyes to the God of the mountains, lift them to the Almighty Dollar.

Dam Hetch Hetchy! As well dam for water-tanks the people's cathedrals and churches, for no holier temple has ever been consecrated by the heart of man.

BIBLIOGRAPHICAL ESSAY

THE CLASSIC ARTICULATION of the conservation movement is Samuel Hays's *Conservation and the Gospel of Efficiency* (Cambridge, Mass.: Harvard University Press, 1959), which is still a must-read for anyone studying conservation or progressivism. Some other older works are also still of value, including Elmo R. Richardson's *The Politics of Conservation: Crusades and Controversies, 1897–1913* (Berkeley: University of California Press, 1962) and James Penick's *Progressive Politics and Conservation* (Chicago: University of Chicago Press, 1968). Harold K. Steen's *The U.S. Forest Service: A History* (Seattle: University of Washington Press, 1976) is an accessible institutional history that covers much more than the Progressive Era; the same is true of John Ise's *Our National Park Policy: A Critical History* (Baltimore, Md.: Johns Hopkins University Press, 1961). For the classic telling of the Hetch Hetchy battle, see Roderick Nash, *Wilderness and the American Mind* (New Haven, Conn.: Yale University Press, 1967). John F. Rieger's *American Sportsmen and the Origins of Conservation* (Corvallis: Oregon State University Press, 2001; originally published 1975) offers a different perspective on the movement, arguing that hunters were primarily responsible for the development of the conservation ethic and the movement. In *The American Conservation Movement: John Muir and His Legacy* (Madison: University of Wisconsin Press, 1981), Stephen Fox casts Muir as a leading figure in the movement.

Reclamation of arid lands has received considerable attention from western historians. See particularly Donald J. Pisani, *To Reclaim a Divided West: Water, Law, and Public Policy, 1848–1902* (Albuquerque: University of New Mexico Press, 1992) and Donald Worster, *Rivers of Empire: Water, Aridity, and the Growth of the American West* (New York: Oxford University Press, 1985). On protecting wildlife, see Thomas R. Dunlap, *Saving America's Wildlife* (Princeton, N.J.: Princeton University Press, 1988).

Much can be learned about the conservation movement through the study of individual lives. Gifford Pinchot authored a lengthy autobiography, *Breaking New Ground* (New York: Harcourt, Brace & Co., 1947), a large part of which recounts the movement from the perspective of Washington, D.C. Although there are other biographies of Pinchot, the most valuable is Char Miller's *Gifford Pinchot and the Making of Modern Environmentalism* (Washington, D.C.: Island Press, 2001).

There are dozens of works concerning Theodore Roosevelt's life, but Paul Cutright's *Theodore Roosevelt: The Making of a Conservationist* (Urbana: University of Illinois Press, 1985) addresses conservation most thoroughly. Cutright grounds Roosevelt's politics in his youthful exposure to and interest in natural history.

The third figure most often closely associated with Progressive Era environmental reform is John Muir. The first complete biography of Muir, Linnie Marsh Wolfe's *Son of the Wilderness: The Life of John Muir* (New York: Knopf, 1945) is still a valuable work. Wolfe praises the heroic Muir as the true voice of conservation. Michael P. Cohen's *The Pathless Way: John Muir and the American Wilderness* (Madison: University of Wisconsin Press, 1984) is a nontraditional biography that focuses on Muir's relationship with the Sierra Mountains. Thurman Wilkins's *John Muir: Apostle of Nature* (Norman: University of Oklahoma Press, 1995) offers a more traditional biography.

Clearly, a focus on biography—long a tradition in conservation history—lends itself to a male-centered, top-down interpretation of the movement. Caroline Merchant played an important early role in revising this dominant interpretation. Frustrated with the male-centered vision of the movement, Merchant authored "Women and the Progressive Conservation Crusade" (*Environmental Review* 8, 1984, 57–85), in which she emphasizes the significant but forgotten role of women.

More recently, Louis S. Warren's *The Hunter's Game: Poachers and Conservationists in Twentieth-Century America* (New Haven, Conn.: Yale University Press, 1997) offers another overlooked perspective on conservation, revealing how regulations, while benign to the environment, could be remarkably imperial. Conservation in this interpretation is an important part of the market expansion into the West, not a reaction against that expansion. Mark Spence has made a similar argument in *Dispossessing the Wilderness: Indian Removal and the Making of National Parks* (New York: Oxford University Press, 1999), a study of Yellowstone, Yosemite, and Glacier National Parks, which reveals the native dispossession required to create these preserves. With *Crimes Against Nature: Squatters, Poachers, Thieves, and the Hidden History of American Conservation* (Berkeley: University of California Press, 2001), Karl Jacoby continues the recent trend of studying conservation from the bottom up. Jacoby is concerned with the consequences of conservation legislation on rural populations, and local actors move to the foreground.

Long ignored by students of conservation, urban environmental reforms have gained considerable attention of late. Of most value are William H. Wilson, *The City Beautiful Movement* (Baltimore, Md.: Johns Hopkins University Press, 1989); Martin Melosi, *The Sanitary City: Urban Infrastructure in America from Colonial Times to the Present* (Baltimore, Md.: Johns Hopkins University Press, 2000); and Suellen Hoy, *Chasing Dirt: The American Pursuit of Cleanliness* (New York: Oxford University Press, 1995). See also David Stradling, *Smokestacks and Progressives: Environmentalists, Engineers, and Air Quality in America, 1881–1951* (Baltimore, Md.: Johns Hopkins University Press, 1999).

On the Progressive Era context of conservation, see Robert Wiebe's classic *The Search for Order, 1877–1920* (New York: Hill & Wang, 1967) and Daniel T. Rodgers's more recent *Atlantic Crossings: Social Politics in a Progressive Age* (Cambridge, Mass.: Belknap Press of Harvard University Press, 1998).

Since the development of modern environmentalism in the 1960s, historians have embarked on an effort to locate the "roots" of the movement. The Progressive Era conservation movement has attracted considerable attention, but recently historians have begun to seek out the origins of conservation itself. Richard H. Grove offers a controversial argument in *Green Imperialism: Colonial Expansion, Tropical Island Edens and the Origins of Environmentalism, 1600–1860* (New York: Cambridge University Press, 1995). Grove found evidence that Europeans were forced to practice conservation as they colonized tropical islands. Richard W. Judd locates the roots of American conservation closer to home, in the lives of average men and women who made their living from the land in New England. In *Common Lands, Common People: The Origins of Conservation in Northern New England* (Cambridge, Mass.: Harvard University Press, 1997), Judd finds evidence of a developing conservation ethic in a land of growing scarcity.

David Lowenthal's recent biography of George Perkins Marsh brings us full circle. In *George Perkins Marsh: Prophet of Conservation* (Seattle: University of Washington Press, 2000), Lowenthal attempts to reinstate Marsh as the central figure in the development of conservation ideals. Marsh was the first to articulate in print a growing concern about humanity's ability to dramatically alter the environment. His *Man and Nature: Physical Geography as Modified by Human Action* was originally published by C. Scribner & Co. in 1864; an edition edited by David Lowenthal was published by Harvard University Press in 1965 and was reissued in 2003 by the University of Washington Press with a new introduction by Lowenthal. As Lowenthal argues, *Man and Nature* became remarkably influential in Europe and the United States, sparking the protection of watersheds and the creation of the forestry division within the federal government.

Conservation clearly did not disappear when the reformism of the Progressive Era subsided. Historians have tracked conservation policies into the 1920s and into the 1930s, when conservation, under FDR, once again became a hallmark of a reform-minded administration. Of particular interest are Kendrick A. Clements, *Hoover, Conservation, and Consumerism: Engineering the Good Life* (Lawrence: University Press of Kansas, 2000) and Donald C. Swain, *Federal Conservation Policy, 1921–1933* (Berkeley: University of California Press, 1963). For an introduction to New Deal conservation, one might begin with A. L. Riesch Owen, *Conservation under F.D.R.* (New York: Praeger, 1983).

INDEX

Addams, Jane, 9, 65
Adirondack Mountains, 4, 53
agriculture, 28–31
air pollution, 73–82; in Cincinnati, 76; in St. Louis, 78–79, 82–83. *See also* smoke
Antiquities Act of 1906, 8
Audubon Society, 6, 49–50

Beard, Mary Ritter: "Civic Improvement," 64–66
beauty: municipal, 64–65; necessary to conservation, 61–63
birds, 47–48, 49–50

campgrounds, 92, 99
Catlin, George, 6
Census of 1890, 4
Chicago, 62
Cincinnati, 62, 75–77
cities: air pollution in, 73–83; improvement of, 64–66; sanitation in, 71; and water, 62–63, 71–72
City Beautiful, 64–65
civic leagues, 64–66
civilization, 10
Cleveland, 62
Cleveland, Grover, 40

coal, 20; ownership of lands, 36; and smoke abatement, 80–83
common good, as principle of conservation, 21–22, 26–27. *See also* public good
conservation: broad applications of, 22; criticisms of, 35–39, 39–41; definition of, 13, 32–33; economic, 59–60; historians' interpretations of, 11–13; importance to nation, 24; Pinchot's definition of, 22; as public duty, 71; utility of, 57–58

democracy: irrigation and, 28; in reform language, 10; supported by conservation, 26; threatened by conservation, 14
development, 20

education, 49–50
efficiency, 22, 26; in coal combustion, 80–83; economic benefits of, 82–83; national, 66–67
engineers, 12; sanitary, 70; and smoke, 80–81, 82
environmental movement, 11
epidemics, 71

Fernow, Bernard, 5

Fisher, Irving: "National Vitality, Its Wastes and Conservation," 66–69

fishing, 53–56

forest fires, 20–21

Forest Management Act of 1897, 5

Forest Reserve Act, 5

forests, 4–5, 19; management, 37, 39–41; protection of, 25, 34

freedom, 26. *See also* democracy; individual rights

future: planning for, 25–26, 33–34, 39, 67–68

game, 45–48, 51–52

garbage, 72

Garfield, James A., 89, 99

Gompers, Samuel: "Conservation of Our Natural Resources," 59–60

government: promotion of public health, 68–69; role in conservation, 33, 36–39, 39–41, 71–72

Grant, Ulysses S., 6

Grinnell, George Bird, 6; "American Game Protection: A Sketch," 45–48

Harrison, Benjamin, 5

Hays, Samuel, 11–12

health: conservation of, 66–69; effects of smoke on, 76–77; linked to prosperity, 66–68

Hetch Hetchy Valley, 13, 85–101; physical description of, 92–94, 95–97, 99–100

Hornaday, William T.: *Our Vanishing Wildlife: Its Extermination and Preservation*, 51–52

Hough, Franklin B., 4

hunters. *See* sportsmen

hunting, 6, 45–46, 51–52, 54–55

hygiene, 68–69

individual rights: to clean air, 77; threatened by conservation, 39

irrigation, 28–31

Keith, William, 99

Knapp, George L.: "The Other Side of Conservation," 35–39

Kroeger, Mrs. Ernest R.: "Smoke Abatement in St. Louis," 78–79

labor, 59–60, 66

Lacey Act of 1900, 8

Ladies' Home Journal: "What Is Meant by Conservation?" 32–35

land, public, 26, 38

logging, 34, 39–41

Marsh, George Perkins, 3–4

Mattes, H.J.M.: "Another National Blunder," 39–41

McFarland, J. Horace: "Shall We Have Ugly Conservation?" 61–63

Merrill, David Shepard: "The Education of a Young Pioneer in the Northern Adirondacks," 53–56

mineral resources, 25, 63.

mining, 63

Minneapolis, 62

Muir, John, 13, 61, 86; "Hetch Hetchy Valley," 94–101

National Conservation Association, 35

National Conservation Commission, 9, 61, 63

National Forests, 38

National Forest Service, 23

Native Americans, 12

natural resources: conservation of, 19–23, 67–68; first inventory of, 24; labor as, 59–60, 66; waste of, 59. *See also* forests; mineral resources; soil; water

nature, control of, 30–31

net fishing, 53–56

Newlands Act, 5

New York State, 46, 53

Oakland, 87

Ohle, Ernest L.: "Smoke Abatement: A Report on Recent Investigations Made at Washington University," 82–83

Olney, Warren, 85; "Water Supply for the Cities About the Bay of San Francisco," 87–90

Parsons, Edward Taylor, 85; "Proposed Destruction of Hetch-Hetchy," 91–94

passenger pigeon, 6

Pelican Island, Florida, 8, 48

Peshtigo, Wisconsin, 5

Philadelphia, 62

Pinchot, Gifford, 8, 11, 85, 89; "Principles of Conservation," 19–23

Pittsburgh, 80

population growth, 25

poverty, 66–67

power plants, 80–83

preservation, 13; contrasted with conservation, 33–34; of game, 45–48, 51–52; of Hetch Hetchy valley, 85–86, 91–101

Progressivism, 9–10

property rights, 21

public good, 25, 26–27, 71–72, 90

race, 67–68

Reed, Charles A. L.: "An Address on the Smoke Problem," 75–77

reforestation, 34

refuges, 48

Richards, Ellen H.: *Conservation by Sanitation: Air and Water Supply, Disposal of Waste*, 70–72

Roosevelt, Franklin, 10

Roosevelt, Theodore, 7, 8, 33, 48, 89; "Special Message from the President of the United States," 23–27

San Francisco, 85, 87–90, 91–93

sanitation, 70–72

Sargent, Charles, 8

sawmills, 40. *See also* logging

scenic landscapes, 61–63, 91–101

science: and agriculture, 30–31; and sanitation, 71–72

Sierra Club, 13–14, 85

Sierra Nevada, 87, 88

smoke, 6, 73–83; symbolism of, 80, 81

Smythe, William E.: "The Miracle of Irrigation," 28–31

soil, 71–72

sportsmen: as conservationists, 45–47, 51–52; publications of, 46–47

St. Louis, 62; smoke abatement in, 78–79, 82–83

suburbs, 65–66
surveys, 65

Taft, William Howard, 9
Tarbell, Ida, 9
technology: and natural resources, 67; and smoke abatement, 80–81
Tuolumne River, 88–89, 90

unemployment, 59–60, 66
utilities, 88
urban planning, 64–66

Van Hise, Charles, 11

waste: definition disputed, 38; prevention of, 20–21, 59; represented by smoke, 81; unemployment as, 59–60, 66

water: in American cities, 62–63, 71–72; drinking, 87–90; quality management, 71–72; used for transportation, 20, 25. *See also* irrigation
Wildlife. *See* game
Williams, Michael, 8
Wilson, Herbert M.: "The Cure for the Smoke Evil," 80–81
Wilson, Woodrow, 86
women: and civic improvement, 64–66; and smoke, 75, 78–79
Women's Organization for Smoke Abatement (St. Louis), 79
Wright, Mabel Osgood: "Keep on Pedaling!" 49–50

Yellowstone National Park, 6
Yosemite National Park, 6, 61; and Hetch Hetchy Valley, 91, 92, 94

WEYERHAEUSER ENVIRONMENTAL BOOKS

The Natural History of Puget Sound Country
by Arthur R. Kruckeberg

Forest Dreams, Forest Nightmares:
The Paradox of Old Growth in the Inland West
by Nancy Langston

Landscapes of Promise: The Oregon Story, 1800-1940
by William G. Robbins

The Dawn of Conservation Diplomacy: U.S.-Canadian
Wildlife Protection Treaties in the Progressive Era
by Kurkpatrick Dorsey

Irrigated Eden: The Making of an Agricultural Landscape
in the American West
by Mark Fiege

Making Salmon: An Environmental History
of the Northwest Fisheries Crisis
by Joseph E. Taylor III

George Perkins Marsh, Prophet of Conservation
by David Lowenthal

Driven Wild: How the Fight against Automobiles
Launched the Modern Wilderness Movement
by Paul S. Sutter

The Rhine: An Eco-Biography, 1815-2000
by Mark Cioc

Where Land and Water Meet:
A Western Landscape Transformed
by Nancy Langston

The Nature of Gold: An Environmental History
of the Klondike Gold Rush
by Kathryn Morse

Faith in Nature: Environmentalism as Religious Quest
by Thomas Dunlap

CYCLE OF FIRE BY STEPHEN J. PYNE

Fire: A Brief History

World Fire: The Culture of Fire on Earth

Vestal Fire: An Environmental History, Told through Fire,
of Europe and Europe's Encounter with the World

Fire in America: A Cultural History of Wildland and Rural Fire

Burning Bush: A Fire History of Australia

The Ice: A Journey to Antarctica

CPSIA information can be obtained at www.ICGtesting.com
Printed in the USA
BVOW070829030412

286706BV00001B/4/P